The Semantics of English Negative Prefixes

Dedication

This work is dedicated to the soul of my mother
(1931-2008)
who planted its seeds, supported its progress
but didn't live to reap its fruits

The Semantics of English Negative Prefixes

Zeki Hamawand

SHEFFIELD UK BRISTOL CT

Published by

UK: Unit S3, Kelham House, 3 Lancaster Street, Sheffield, S3 8AF

US: ISD, 70 Enterprise Drive, Bristol, CT 06010

www.equinoxpub.com

First published in hardback in 2009. This paperback edition published in 2013.

British Library Cataloguing-in-Publication Data
A catalogue record for this book is available from the British Library.

ISBN-13	978-1-84553-540-7	(hardback)
	978-1-84553-971-9	(paperback)

Library of Congress Cataloging-in-Publication Data

Hamawand, Zeki.
The semantics of English negative prefixes / Zeki Hamawand.
p. cm.
Includes bibliographical references and indexes.
ISBN 978-1-84553-540-7
1. English language--Negatives. I. Title.
PE1359.N44H36 2009
425--dc22
2008046893

Typeset by Catchline, Milton Keynes (www.catchline.com)
Printed and bound in Great Britain by Lightning Source UK Ltd., Milton Keynes and Lightning Source Inc., La Vergne, TN

Contents

List of figures

List of tables

Preface

Abstract

This study proposes a new system for describing the semantics of negative prefixes in English. Theoretically, the study builds on assumptions of Cognitive Semantics. One assumption is that lexical items form complex categories. Applied to negation, it seeks to show that negative prefixes form categories of interrelated senses arranged with respect to a primary sense. Another assumption is that lexical items gather in cognitive domains. Extended to negation, it strives to show that negative prefixes form sets within which they name different facets. A further assumption is that lexical items display non-synonymy. Related to negation, it attempts to show that prefixally-negated pairs are different although they share the same bases. Empirically, the study hinges on assumptions of Usage-based Semantics. One assumption is that knowledge of language is derived from language use. In virtue of this, it tries to link the speaker's knowledge of the prefixes to their situated instances of use. Another assumption is that the linguistic system is shaped by actual data. In view of this, it tries to conduct the analysis of prefixes on authentic data driven from the corpus and/or Internet pages. A further assumption is that form and meaning are tightly linked. In light of this, it aims to indicate that each prefixally-negated word is associated with a special meaning. To disambiguate meanings, it uses discriminating collocates which are associated with the pairs.

Layout

Although much has been written on individual negative prefixes, the present study is unique in treating them as a coherent class. Chapter 1 develops the semantic system needed to address the question of forming negative words by means of prefixes. It introduces the three semantic axes which account for the analysis of the negative prefixes. Chapter 2 gives an outline of the area of derivational morphology, laying emphasis on the conceptions involved in its description. Chapter 3 introduces the first axis in my semantic system, which focuses on the question of category in the semantic characterisation of negative prefixes. Category is the study of the syntagmatics of word structure, whereby a negative prefix is combined with a base to create a new word. Chapter 4 introduces the second axis in my semantic system, which concentrates on the question of domain in the semantic characterisation of negative prefixes.

Domain is the study of the paradigmatics of word structure, whereby a number of negative prefixes stand for the same concept. Chapter 5 introduces the third axis in my semantic system, which explores the question of construal in the semantic characterisation of negative prefixes. Construal is the study of the interpretation of word structure, whereby pairs of words sharing same bases are distinguished according to semantic principles. Chapter 6 looks back and summarises in text and tabular form what the study has accomplished.

Acknowledgements

This book represents research into the nature of the English lexicon, providing a comprehensive elucidation of negative prefixes. In the course of preparing the book, a number of people have contributed directly or indirectly to its evolution. Firstly, I would like to thank the following linguist colleagues for helpful comments on various aspects of the work: Ronald Langacker of the University of San Diego, Lynne Murphy of the University of Sussex, Susan Hunston of the University of Birmingham, Julie Coleman of the University of Leicester, Gerald Nelson of University College London and Nathalie Schapansky of the University of Alberta. In fact, I have benefited greatly from their feedback. Secondly, I owe a special debt to Wendy Anderson, of the University of Glasgow, who read through the manuscript and provided me with invaluable suggestions, all of which have been incorporated in the study. Thirdly, I am much grateful to Laurence R. Horn, of Yale University, and Royal J. Skousen, of Brigham Young University, who sent me copies of articles from which this study has greatly benefited. Finally, my thanks go to the team at Equinox, Janet Joyce, David Graddol and Val Hall, for overseeing the review of the manuscript and putting it into its final shape.

1 Negation

1.0 Overview

This chapter introduces a new semantic system that is capable of justifying the formation of negative words via prefixation, a morphological process whereby a bound morpheme is attached to the front of a base, as in *defrost*. To tackle the issue of negative prefixation, the chapter is divided into five sections. Section 1 presents an outline of the issue, specifying its theme, context and questions. Section 2 reviews previous endeavours on the issue. The section is subdivided into three parts. Part 1 sets out the way language dictionaries treat the issue. Part 2 describes how reference grammars deal with the issue. Part 3 investigates the way morphological studies tackle the issue. Section 3 proposes a new way of tackling the issue, focusing on its claims, goals and steps. Of particular interest is the capability of the system to account for all sorts of complexities in the process of deriving negative words. Section 4 elaborates on the methodology: part 1 touches upon tenets of the theoretical method, namely Cognitive Semantics, and part 2 brings up tenets of the empirical method, namely Usage-based Semantics. Section 5 looks back on the chapter as a whole and sums up its content.

1.1 Introduction

Research theme. This study is about *negation* in English. Simply defined, *negation* refers to a morphosyntactic operation in which a lexical item is used to deny the truth of an expression. As the definition reveals, negation is of two types: syntactic and morphological.[1] Syntactic negation is the process of negating an expression by using negators like *no*, *not* or *never*, as in *She is not happy*, or words having negative senses like *hardly*, *rarely* or *scarcely*, as in *There is scarcely any coffee left*. Morphological negation is the process of negating an expression by adding affixes to bases. This type of negation is difficult to describe as it covers diverse processes. Affixes in English are of two sorts: prefixes and suffixes. Negative prefixes are lexical items that are added to the beginnings of bases to form words, as in *unhappy*. Negative suffixes are lexical items that are added to the ends of bases to form words, as in *cordless*. Of the two sorts of affixal negation, the discussion here focuses on the semantic description of negative prefixes. English provides its speakers with a variety of such prefixes including *a-, ab-, anti-, contra-, counter-, de-, dis-, in-, mal-, mis-, non-, pseudo-, quasi-, semi-, sub-, un-* and *under-*.[2]

Research context. A look at the data shows that negative prefixes display three linguistic tendencies. First, negative prefixes often attach to bases of different word classes. For example, in *non-cooperation* the prefix *non-* is attached to a noun. In *non-essential*, it is attached to an adjective. In *non-skid*, it is attached to a verb. In such occurrences, the prefix has been regarded as monosemous in the literature, i.e. having a single meaning. Second, negative prefixes sometimes express more or less the same meanings. For example, in *defog*, *dispossess* and *unchain* the prefixes *de-*, *dis-* and *un-* are used to denote the concept of *removal*. In such uses, the prefixes have been regarded as synonymous in the literature, i.e. expressing identical meaning. Third, negative prefixes occasionally attach to the same bases. For example, the words *dissatisfied* and *unsatisfied* are both derived from the same base *satisfied*. Owing to their morphological relationship, such pairs have typically been regarded as free variants in the literature, i.e. substitutable for one another without a change in meaning. The scope of the present study concerns itself with the semantic properties of negative prefixes as shown in such actual patterns.[3]

Research questions. The investigation of the occurrences of the negative prefixes gives rise to three central questions, which are part of the key issue of characterising the nature of the lexicon in general and the structure of complex words in particular.

- Does a negative prefix exhibit multiple senses, and if so, on what basis are its senses organised? Do the senses derive from a primary sense, and if so, how is the primary sense identified?
- Do negative prefixes form semantic sets, and if so, on what basis are they grouped together? Do they represent different facets within the sets, and if so, how do they contrast with one another?
- Do pairs of words derived by negative prefixes have different readings, and if so, in what respect are they different? Is the difference supported by evidence, and if so, where does it come from?

System chosen. To answer the questions above, a new system of morphology is needed. The new system has to address three linguistic phenomena. One phenomenon pertains to *lexical multiplicity*, whereby negative prefixes display a series of senses which gather around a nucleus. The senses are organised in terms of distance from the nucleus, based on the degree of similarity. Another phenomenon revolves around *lexical relationship*, whereby negative prefixes cluster in sets defined by two types of relation: one is of similarity vis-à-vis the overall concept of the set they form; the other is of difference with respect to the specific functions they perform within the set. A further phenomenon relates to *lexical alternation*, whereby two, or more, negative prefixes compete to derive new forms from the same bases. The resulting forms are considered alternatives, exhibiting both phonological distinctness and semantic dissimilarity. In spite of sharing the same bases, each alternative has a distinct function to carry out in the language. The viability of the new system rests on the tenets of Cognitive Semantics and Usage-based Semantics, both of which give meaning a central position in language.

Reasons for choice. The new system is chosen because it is adept at handling analyses of messy arrays of linguistic data, which are shunned by other theories. Related to the present theme, it is assumed that it will present a cogent treatment of the three lexical phenomena which have not yet been explained in minute detail or covered in a comprehensive manner in the literature. Cognitive Semantics has been chosen because it allows one to study the intricacies of linguistic structures in a coherent fashion. It lays emphasis on explaining linguistic structures with reference to cognitive processes. It claims that cognition and language influence each other. Both are embodied in the experiences of language users. Cognition triggers language, whereas language is processed by cognition. Usage-based Semantics has been chosen because it allows one to probe the full range of language use, regular and irregular. It focuses on actual patterns of linguistic structures. It claims that knowledge of language arises out of language use and is influenced by context, the information surrounding a word which helps make its meaning clear.

Before introducing the details of the new system, let us first survey the literature and see how such questions are addressed by the previous endeavours in language study.

1.2 Previous endeavours

The area of negative prefixation in English has long been of concern to linguists interested in the morphology of the language.[4] The process of deriving a single word from a base has been treated in some detail. Some studies like Algeo (1971), Maynor (1979), Colen (1980), Thomas (1983), Andrews (1986) and Horn (2002) deal with individual negative prefixes. Other studies like Marchand (1969), Lehrer (1995), Urdang (1982), Adams (2001), Plag (2003) and Lieber (2004) survey a selection of negative prefixes. Because these studies address the issues of lexicalisation, productivity and semantic drift, the current project will place these issues outside its scope. By contrast, very little has been written on the process of deriving a pair of words from a base, i.e. cases where the same base accepts two negative prefixes. Even major studies like Zimmer

(1964), Funk (1971) and Horn (1989) leave crucial facts regarding such cases out of consideration. No doubt, the literature contributes to our understanding of word formation, but it provides no in-depth or exhaustive treatment of negative prefixation. One reason for this is that linguists refrain from talking about such an area because it is fraught with complications. Another, more important, reason is that linguists lack the necessary tools to investigate such a significant topic in morphology.

To get a clear picture of previous treatments, I shall briefly review the literature. First, I classify the literature into different types of endeavour, dealing with each in turn: section 1 is devoted to lexicographical endeavours; section 2 is a discussion of grammatical endeavours; and section 3 is a review of morphological endeavours. Second, I single out the endeavours that are the most prominent or influential on the topic, and examine the central claim of each endeavour. Third, I assess each type of endeavour with reference to the three lexical phenomena: a negative prefix can convey different meanings; a single concept can be expressed by different negative prefixes; and alternative pairs exhibit semantic distinctions. The purpose of the review is not to criticise any of the endeavours, but to see what they can bring to the current analysis or to see if the present argument can build on any of them. These endeavours are indispensable references for any research in the analysis of English morphology. They bring interesting insights to the study of language in general although they present some limitations with reference to the specific topics which relate to negative prefixation.

1.2.1 Lexicographical endeavours

My first review of the literature covers language dictionaries. Most dictionaries of language, e.g. COBUILD (2001), LDOCE (2003) and OALD (2005), contain a list of lexical items in alphabetical order with information about their pronunciation, meaning and register. However, perusing the dictionaries for answers to the central questions shows that they display considerable deficiencies. Firstly, dictionaries give only a very sketchy account of the negative prefixes under investigation. In this way, dictionaries overlook some important senses of the negative prefixes. By making such vague descriptions, dictionaries neglect to present a full picture of the negative prefixes. Secondly, dictionaries describe the lexicon by allotting the prefixes separate entries. In this way, dictionaries fail to show that these prefixes have something in common as well as something to distinguish them. Dictionaries stop short of providing accurate definitions, practical examples or usage notes to highlight the differences. Thirdly, dictionaries present pairs of words sharing common bases as semantically interchangeable. In this way, dictionaries disregard the fact that every word has a separate message to convey in the language. Dictionaries fail to deal with the choice between the pairs or hint at the factors that determine their selection. In sum, the types of information that dictionaries provide are not entirely pertinent to the needs of the language users.[5]

1.2.2 Grammatical endeavours

My second review of the literature covers reference grammars. Most grammars of language, e.g. Quirk et al. (1985), Biber et al. (2002) and Huddleston & Pullum (2002), offer an array of rules that define the structure of the language. Nevertheless, scanning the grammars for answers to the central questions shows that they have some shortcomings. One shortcoming is that these grammars present only a superficial account of the negative prefixes, which is limited to a mere listing of their various types. Detailed descriptions of the multiple meanings of the negative prefixes are almost non-existent. Without such notes, the language user would not be able to see the exact behaviour of any prefix. Another shortcoming is that these grammars make no reference to alternatives that are related to a target prefix. A considerable amount of necessary data concerning important usages of the prefixes is missing. Without such notes, the language user would not be able to locate the alternative prefixes that exist in the language. A further shortcoming is that these grammars rarely differentiate between the resulting pairs, nor do they provide information regarding the use of each option. They do not offer guidance for making a distinction in meaning between such pairs. Without such notes, the language user would not be able to choose the appropriate word in discourse. In brief, such a state of affairs surely blurs the uses of the alternative prefixes in question.

1.2.3 Morphological endeavours

My final review of the literature includes work on morphology. Prefixal negation has been dealt with by linguists in the major linguistic schools. Works of morphology, e.g. Zimmer (1964), Funk (1971) and Horn (1989), deal with the structure and form of negative words. Nonetheless, scrutinising the works in search of answers to the central questions shows that they suffer some inadequacies. One inadequacy is that although a great part of their investigation is dedicated to the semantic restrictions on the use of the negative prefixes, they make no apparent contribution to the multiple readings of any prefix. In doing so, they miss drawing a fully-articulated account for each prefix. Another inadequacy is that while these works present coverage of the negative prefixes, they refrain from treating them as groups or giving cross references to their counterparts. In doing so, they fail to reveal the distinctive roles which the prefixes play in the language. A further inadequacy is that although these works show an awareness of the existence of alternative words, they don't capture the differences in meaning between them in a systematic way. Typically, they tackle a few pairs of words, but do not provide the parameters for differentiating between them. In this way, they do not take into account the role of alternative prefixes in triggering differences in the meanings of the alternative derivatives. In short, these morphological works overlook the fact that alternation is a widespread phenomenon in language.

From the preceding discussions, a few conclusions emerge. Although the foregoing endeavours offer some useful hints on the topic of negative prefixation, they are deficient when it comes to a fuller explanation. The lexicographical endeavours, which have been predominantly corpus-based in nature, have yielded little success in accounting for the differences in meaning between prefixally-negated alternative pairs. The grammatical endeavours have not provided a unified explanation for the full array of semantic properties of the negative prefixes. The morphological endeavours have offered little for our understanding of the lexical relationships between the negative prefixes. In sum, although these endeavours differ in scope and presentation, they are alike in inadequately dealing with the semantics of negation in word formation. To remedy these deficiencies, it is reasonable to seek a more satisfactory explanation for negative prefixation. In other words, in order to gain a better insight into the ways in which language is structured, a new system with the right properties is needed, a system in which various semantic axes interact with the aim of providing an accurate picture of the behaviour of the lexical items under investigation.

1.3 A new endeavour

To come to grips with the topic of negative prefixation, the new system which I suggest is an extension of my own previous work on word structure (Hamawand, 2007, 2008). In this system, morphology is the study of words in which shape adjustments, triggered by cognitive processes, reflect meaning alterations. One such process is prefixation, whereby a dependent morpheme (prefix) is integrated, due to phonological and semantic correspondences, with an autonomous morpheme (base) to form a complex structure (new word). To provide a plausible solution, the present endeavour couples together mechanisms of two linguistic methods: Cognitive Semantics and Usage-based Semantics. These methods are essential for three reasons. First, they help one to explore the subtle differentiations in the multiple definitions of any negative prefix. Second, they help one to understand the specific meanings of the negative prefixes by grouping them in sets. Third, they help one to explain the cases where alternatives are derived from the same bases by means of two or more negative prefixes. Although the prefix variants are admissible, each assigns a different meaning to the host base. Section 4 elaborates on the two linguistic methods.

1.3.1 Axes

To answer the questions at the heart of the present endeavour, my semantic system for explaining the formation of negative words rests on three semantic axes. The first pertains to the profiles of the negative prefixes. In this respect, I claim that a negative prefix has a multiplicity of senses which gather around a core. It is here that I initiate multiple definitions for each negative prefix and concentrate on its lexical properties. The second axis concerns the clusters of the negative prefixes. In this regard, I claim that, based on similarities in their definitions, negative prefixes group together in conceptual sets. It is here that I discuss the semantic characteristics of the sets in which the prefixes group, and then establish the specificity of each prefix when contrasted with a counterpart within a given set. The third axis relates to the interpretations of the derivatives. In this connection, I claim that alternations of prefixes always have semantic consequences, and the alternatives they form are in no way interchangeable although they share the same base. Here, I posit the semantic parameters that separate the alternatives and impart to each alternative a specialised role in the language.

The choice of an alternative, I claim, is a matter of two factors. One factor resides in *conceptual content*, the meaning conventionally associated with an expression. The content of a base is multi-faceted, whereas a prefix has its own content. In the combination process, the prefix imposes its content on that of the base, and so gives the resulting alternative a different meaning. That is, when two alternative prefixes attach to the same base, each serves to highlight a different facet of the base's content. Each of the resulting alternatives encodes, therefore, a distinct meaning. Another factor resides in *construal*, the way the content is conceived relative to the communicative needs. The meaning of a derived alternative involves the particular construal the speaker employs to describe a situation. Linguistically, the construal is encoded by means of a prefix. That is, when two negative alternatives have the same conceptual content, they differ semantically by virtue of the construals they represent and linguistically by the negative prefixes they host. The negative prefixes are, therefore, responsible for separating the alternatives.

By way of illustration, let us consider the pair below and see how it is dealt with by the different endeavours:

- dissatisfied vs. unsatisfied
- The guest was *dissatisfied* with the service at the hotel.
- They reported an *unsatisfied* demand for fresh product.

Let us first examine lexicographical endeavours. In COBUILD, the entry for *dissatisfied* (2001: 443) is 'if you are dissatisfied with something, you are not contented or pleased with it', whereas the entry for *unsatisfied* (2001: 1717) includes both 'if a need or demand is unsatisfied, it is not dealt with' and 'if you are unsatisfied with something, you are disappointed because you have not got what you hoped to get'. In LDOCE, the entry for *dissatisfied* (2003: 452) is 'not satisfied because something is not as good as you had expected', whereas the entry for *unsatisfied* (2003: 1815) includes both 'an unsatisfied demand, request, etc. has not been dealt with' and 'not pleased because you want something to be better'. In OALD, the entry for *dissatisfied* (2005: 364) is 'not happy or satisfied with something', whereas the entry for *unsatisfied* (2005: 1424) includes both '(of a need, demand, etc.) not dealt with in a satisfying way' and '(of a person) not having got what you hoped; not having had enough of something'. A look at these attempts leads one to conclude that (i) the definitions are imprecise, (ii) the semantic differentiation is unprincipled, (iii) the examples are not distinctive, (iv) the roles of the prefixes are ignored, and (v) empirical evidence is missing.

Let us secondly inspect grammatical endeavours. In Quirk et al. (1985: 1540–1) the pair under investigation does not get a mention. However, in a note the authors very briefly touch upon the question of alternation. First, they speak of lexicalisation: a word becomes established and is no longer produced according to rules. *Unreplaceable* involves less lexicalisation than *irreplaceable*. Then, they speak of a difference in meaning. *Non-scientific* expresses a contradictory distinction, whereas *unscientific* expresses a contrary distinction. In Biber et al. (2002: 239–48), apart from lists of prefixes, there is no mention of the topic of alternation at all. In Huddleston & Pullum (2002: 1688), a couple of words including the ones under investigation receive a sketchy explanation. In some cases, there is no difference in meaning, as in *in/unadvisable*. In other cases, there is a difference in meaning. *Unsatisfied* means 'not satisfied, unfulfilled'. It characteristically applies to abstract entities. *Dissatisfied* means 'discontented'. It generally applies to humans. A scan of these attempts results in the following conclusions: (i) the explanations are unequivocal, (ii) the distinctions are unpredictable, (iii) the prefixes are acknowledged, (iv) the generalisations are invalid, and (v) the words are indiscriminately exchangeable.

Let us thirdly scrutinise morphological endeavours. Zimmer (1964) provides a subtle analysis of negative prefixation in light of restrictions on the formation of adjectival antonyms. He devotes a great amount of space to the

English prefix *un-*, explaining the process of *un-* derivation and the restrictions imposed on *un-* adjectives. Unfortunately, he does not include examples of pairs derived from the same base to support his analysis. Funk (1971: 377) talks about alternation in two ways. In some examples, the derivatives denote the same idea. He cites *dissatisfactory* and *unsatisfactory*, which he considers synonymous. In other examples, the derivatives differ in lexicalisation. *Incompetent* is lexicalised, whereas *non-competent* is not. Horn (1989: 281) reports on what linguists like Jespersen and Marchand wrote about the subject. He does not refer to the two words in his report, but mentions a couple of others. Instead of explaining semantic contrast, he provides a test of gradability. Adjectives in *un-* or *in-* are gradable, while those in *non-* are not. A consideration of these attempts shows that (i) no prefix contribution is given, (ii) no criterion is used, (iii) no listing of pairs is offered, (iv) no systematic treatment is posited, and (v) no unanimity is reached.

In the semantic system I propose here, the two words are derived from the adjective *satisfied*, whose conceptual content is 'pleased to get what one wants'. In spite of having the same base, they differ in terms of construal. When the speaker has someone as the target of conceptualisation, s/he uses the prefix *dis-* whose conceptual content is 'lacking the thing signified by the base'. In *The guest was dissatisfied with the service at the hotel*, the word *dissatisfied* means 'feeling that something is not as good as it should be', i.e. something has been done but is of poor quality. This meaning is supported by accompanying collocations. The word is preceded by nouns denoting people such as *customer, guest, person, supporter, visitor*; linking verbs such as *appear, become, feel, look, sound*; followed by *with* plus nouns denoting product such as *account, merchandise, product, service, work*, etc. The negative prefix *dis-* is used mainly to describe an animate entity. As evidence, none of the collocates of *dissatisfied* is compatible with *unsatisfied*. For example, it would not be possible to say *unsatisfied wisher* because *wisher* is an animate entity.

When the speaker has something as the target of conceptualisation, s/he uses the prefix *un-* whose conceptual content is 'bereft of what is specified by the base'. In *They reported an unsatisfied demand for fresh product*, the word *unsatisfied* means 'the feeling that something has not been fulfilled as expected', i.e. something is left undone. This meaning is borne out by accompanying collocations. The collocations include nouns denoting need such as *demands, needs, offers, requests, requirements*; or desire such as *cravings, desires, impulses, longings, wishes*; preceding verbs denoting movement such as *go, leave, reflect, remain, return*, etc. The negative prefix *un-* is used mostly to describe an inanimate entity. As evidence, none of the collocates of *unsatisfied* is compatible with *dissatisfied*. For example, it would not be possible to say *a dissatisfied wish* because *wish* is an inanimate entity. From the analysis conducted above, it appears that in each derivation the prefix is considered instrumental because it lends its character to the whole outcome. Each prefix stands for the particular construal which the speaker imposes on the content of the base.

1.3.2 Goals

The present study represents an attempt to devise a new system according to which the questions raised at the beginning of the study can be answered. The questions revolve around prefixal negation, a puzzling area in the study of complex structures in English. In this regard, the study has set itself three principal goals.

The first goal is to substantiate the claim that a given negative prefix has multiple senses comprehended in terms of a *category*, a network consisting of a wide range of senses linked via categorising relationships. In terms of the *category* theory, the senses of a negative prefix gather around a prototypical one, from which the peripheral ones are derived. The senses do not share all the properties of the prototype, and so exhibit minimal differences. This characterisation accords with the cognitive assumption that the meaning of a lexical item is not fixed. Through the creativity of the language user, it can be extended into new realms of experience, thus resulting in new senses. Chapter 3 provides strict definitions for the multiple senses of the negative prefixes under investigation. This would help to avoid the error of considering the meanings of the negative prefixes as homonymous, which is prevalent in past work on the subject.

The second goal is to validate the claim that the meanings of negative prefixes are grasped in terms of a *domain*, a coherent area of conceptualisation relative to which linguistic structures can be characterised. In light of the *domain* theory, the core meaning of a negative prefix is best identified first by linking it to the cognitive context in which it is embedded and second by contrasting it with the other prefixes in the same domain. This characterisation complies with the cognitive assumption that the meaning of a linguistic expression cannot be described solely in terms of a bundle of semantic primitives, but instead emerges from highly structured background of knowledge. Chapter 4 groups the negative prefixes, based on the new definitions, into domains which represent different conceptual structures. This helps to avoid the misconception of favouring treatment of prefixes in isolation, which is commonplace in all previous studies.

The third goal is to vindicate the claim that prefixally-negated expressions differ with respect to *construal*, the way a situation is perceived and conceived. Typically, a situation can be construed in alternate ways. In view of the *construal* theory, variants of form are thought of as alternate possibilities licensed by the linguistic system. The selection of a given variant is governed by the particular construal the speaker imposes on a situation. This characterisation squares with the cognitive assumption that meaning is a prime shaper of morphological structure, which resides in the particular way in which the speaker chooses to describe a scene. Chapter 5 provides the semantic denominators that account for the distinction between alternative pairs of negative expressions. This helps to avoid the fallacy that the pairs are synonymous or free variants, as the majority of earlier accounts mistakenly assumed.

1.3.3 Steps

The present study is a synchronic description of complex words in the formation of which negative prefixes combine with various bases. To retrieve the data, the study uses two types of sources. One source is the *corpus*, a collection of actual language production, recorded utterances or written texts, realised by native speakers. The corpus used here is the *British National Corpus*. To generate concordances of the data, the study uses WordSmith Tools. No corpus, no matter how large it is, can capture the entirety of actual language use, and so the study draws on another source too. The other source is the Internet, a huge dataset which includes countless texts available for searching and examination. In Jewell's (2001: 18) words, the Internet is considered as an ideal source for describing language. For one, it has a wide range of content. There are millions of websites on the Internet, which users can access easily. For another, it offers natural data. There is a large number of highly unedited texts, which are reflective of real language use. To generate hits, the study uses the search engine Google. To accomplish the linguistic mission, the study takes three steps.

The first step attains the goal of *category*. At this step, a categorial sketch for each of the negative prefixes is drawn. The prefixes comprise *a-*, *ab-*, *anti-*, *contra-*, *counter-*, *de-*, *dis-*, *in-*, *mal-*, *mis-*, *non-*, *pseudo-*, *quasi-*, *semi-*, *sub-*, *un-* and *under-*. The sketch involves (i) identifying the primary sense of each prefix, and (ii) pinpointing the multiple senses that derive from it. For each sense, a precise definition is given. This is done by analysing the contexts in which each of the negative prefixes occurs. To support the definitions, three examples of words with paraphrases are given. In the course of constructing the definitions, major reference works on word formation such as Marchand (1969), Urdang (1982) and *Collins COBUILD: Word Formation* (1993) are consulted. Proceeding in this way, the researcher can discover in general the semantic relations that hold among the senses of each of the negative prefixes, and establish in particular the minimal differences that exist among them.

The second step realises the goal of *domain*. At this step, the negative prefixes are grouped together in conceptual sets. A complete account of a set includes (i) specifying membership which is based on the definitions of the prefixes, and (ii) explaining the internal structure which assigns a different meaning to each prefix. The meaning of a prefix arises from its relations of similarity and contrast with other prefixes in the set. To support the subtle nuances of the member prefixes, pairs of words sharing the same bases are picked out. This is done by browsing through the occurrences of the negative prefixes. In the course of collecting the pairs, major manuals on English usage such as Partridge (1961), Greenbaum & Whitcut (1988), Fowler (1996) and Peters (2004) are consulted. Proceeding in this way, the researcher can establish that the prefixes compared have different meanings, and mark the semantic territories covered by them.

The third step actualises the goal of *construal*. At this step, pairs of sentences demonstrating the uses of the alternative words negated by the prefixes are provided. This is done by relying on the corpus, Internet and dictionaries. To make the sentences user-friendly, some are made more concise. For the definitions of the bases of the pairs, such major English dictionaries as COBUILD (2001), LDOCE (2003) and OALD (2005) are used. Then the pairs are distinguished. This is done by analysing their contextual preferences and providing collocational evidence. In sense discrimination, the study builds on the work of Kennedy (1991), Clear (1994), Biber et al. (1998), Kilgarriff & Tugwell (2001) and Williams (2002). However, it departs from them in two ways. First, it looks at pairs that share the same bases but begin with different prefixes, whereas the authors mentioned look at individual or separate words. Second, it places the focus on discriminating collocations, while previous authors use statistics to measure frequencies. Proceeding in this way, the researcher can establish how such pairs are distinctive in use.

1.4 Methodology

The task of the current study is to formulate a theory of negation which is grounded in semantics. Specifically, the task is to present a new way of accounting for alternations between negative prefixes when they are attached to the same bases. To achieve this goal, the study draws on insights from two linguistic methods. One is theoretical; the other empirical. The theoretical approach, represented by Cognitive Semantics, provides linguists with the necessary assumptions. One prominent assumption is the idea of explaining linguistic structures with reference to cognitive processes. The empirical approach, represented by Usage-based Semantics, provides linguists with research tools to verify the assumptions. At the centre of the study is a focus on actual patterns of usage. It is assumed that the two methods can work together and give a cogent description of the way language works. Clearly, it is the linguist who interprets the data, represented by the collocations, discovers the semantic distinctions between the derived pairs and then makes the generalisations.

1.4.1 Theoretical method

Theoretically, the study of prefixal negation is couched in Cognitive Semantics, exemplified by linguists such as Fillmore (1977, 1982), Talmy (1983, 1985), Fauconnier (1985, 1997), Lakoff (1987, 1990) and Langacker (1988a, 1997). Cognitive Semantics is built on four central assumptions. First, conceptual structure is embodied. The nature of conceptual organisation arises from sensory interaction with the external world. Second, semantic structure equals conceptual structure. The meaning of an expression does not refer to an entity in the real world, but to a concept in the mind of the speaker, which is perceptually grounded and experientially based. Third, meaning representation is encyclopaedic. The meaning of an expression subsumes vast repositories of knowledge, both linguistic and non-linguistic. Fourth, meaning construction is conceptualisation. Meaning construction is a dynamic process whereby linguistic units serve as prompts for an array of conceptual operations. The following paragraphs give summaries of three crucial theories of meaning employed in Cognitive Semantics, which apply to the area of prefixal negation.

Lexical items form categories of interrelated senses. Cognitive Semantics, as demonstrated by Lakoff (1987) and Taylor (1989), builds lexical descriptions on the *category* theory, which was first developed by Rosch (1977, 1978). In view of this theory, a lexical item forms a complex category of overlapping semantic senses. The category contains peripheral zones situated around a conceptual centre. The senses gain membership in the category based on similarity rather than identity. The peripheral zones need not have all of the attributes of the centre. That is, they may not conform rigidly to the conceptual centre. Some of the attributes of the centre may appear to be optional at the periphery. The conceptual centre of a category, termed the *prototype*, is the most representative or most salient instance of the category. The periphery of a category includes the remaining instances which are derived from the centre via semantic extensions. For instance, a *kitchen chair* is regarded as the prototype of the *chair* category because it possesses almost all of its features, whereas *rocking chair*, *swivel chair*, *armchair*, *wheelchair* or *highchair* are regarded as being on the periphery because they possess only some of those features.

Let us apply this to our topic. I argue that the negative prefix *ab-*, from Latin, forms a category of distinct senses. The different senses of the prefix arise from the nature of the combining bases. Prototypically, the prefix *ab-* is added to an adjective to express an abstract entity. It means 'move away from the state expressed by the base'. For instance, *abnormal* is the state of being not normal. Extended from this is a sense that describes a physical entity. It means 'locate away from the thing expressed by the base'. For instance, *aboral* means being located away from the mouth. Peripherally, the negative prefix *ab-* is added to a verb to express an action. In this case, it has two senses. One sense is 'wrongly perform the action expressed by the base'. For instance, *abuse* is the action of treating someone wrongly. Another sense is 'release one from the action expressed by the base'. For instance, *absolve* is the action of releasing one from one's obligations. Chapter 3 discloses the full details of this assumption and employs it in the categorisation of the negative prefixes. In the course of the categorisation, some prefixes appear to express more or less the same senses. On the basis of such converging senses, these prefixes will be grouped into sets, referred to as *domains*. It is within these domains that these prefixes can act as alternatives, where they stand for one concept but differ in the specifics. This cognitive assumption will be the subject matter of the next paragraph.

Figure 1 presents a graphical representation which captures the multiple senses of the negative prefix *ab-*. Note that the solid arrow represents the prototypical sense, whereas the dashed arrows represent the semantic extensions.

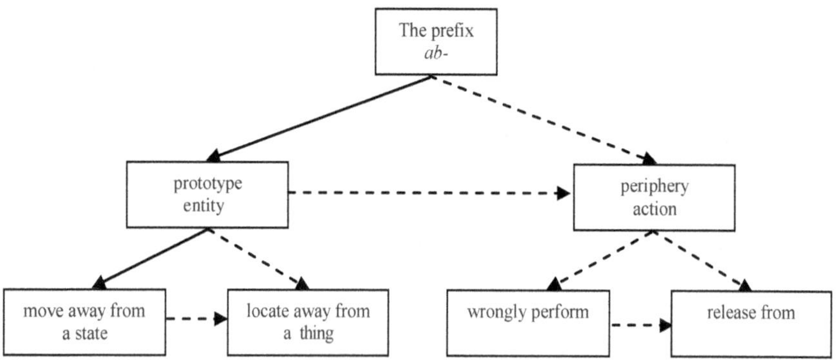

Figure 1: The semantic network of the negative prefix *ab-*

Lexical items gather together in cognitive domains. Cognitive Semantics, as exemplified by Fillmore (1977, 1982) and Langacker (1987, 1991)[6] characterises lexical relationships in terms of the *domain* theory. In light of this theory, the meaning of a lexical item cannot be understood independently of the domain with which it is associated. A *domain* is a knowledge structure or a conceptual entity with respect to which the meaning of a lexical item can be described. A *domain* relates lexical items associated with a scene, situation or event from human experience. A *domain* contains a set of lexical items, each of which is assigned a specific property. To understand the meaning of an item, it is necessary to contrast it with the other items in the same domain. The lexical items which belong to a domain are not in complementary distribution. A close investigation of their behaviour makes it clear they have distinct meanings. The use of each item in language is defined in terms of the background information provided by the domain. For example, the exact meaning of the word *uncle* cannot be identified without activating the domain of *kinship* as the background knowledge for its description.

Let us apply this to our topic. The negative prefixes *ab-*, *sub-,* and *under-* evoke, I argue, the domain of *degradation*, an area of knowledge which entails a decline in rank, importance or size of an entity. As the definition discloses, *degradation* has three facets. The first is quality and is represented by *ab-*. *An abnormal person* is a person who deviates from what is normal. The second is degree and is represented by *sub-*. *A sub-human treatment* is a treatment that is less than human. The third is rank and is represented by *under-*. *An under-servant* is a servant of inferior or subordinate rank. Chapter 4 gives more details about this assumption and utilises it in the classification of the negative prefixes. Prefixes standing as alternatives construe a situation in different ways. It is left to the speaker to decide when to use a particular prefix. The choice of the speaker comes under the rubric of *construal*. The elaboration of this cognitive assumption will be the task of the following paragraph.

Figure 2 presents a configuration which captures the different facets of the domain of *degradation.*

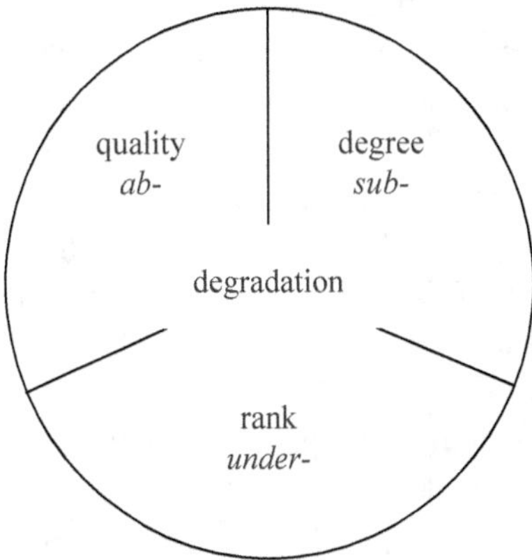

Figure 2: The domain of degradation

Lexical items embody alternative construals of content. Cognitive Semantics, as demonstrated by Langacker (1987, 1991) describes utterance interpretation according to the *construal* theory, whereby the meaning of a linguistic utterance is identified in terms of the way it is construed. The meaning of a lexical item includes both conceptual content and the particular ways of construing that content. *Construal* is the ability of the speaker to conceptualise a situation differently and use different linguistic expressions to represent these different conceptualisations in discourse. Two lexical items may share the same content, but differ in terms of the alternate ways the speaker construes their common content. In each alternative, the speaker adjusts his/her conceptualisation and focuses on a particular aspect of the situation s/he describes. Each alternative is realised in language differently. The constructions *He sent a letter to Susan*, and *He sent Susan a letter* are truth-conditionally equivalent, but they are construed differently. In the prepositional construction, the focus is on the issue of movement that is focused, whereas in the ditransitive construction the focus is on the result of the action that is focused. Therefore, only the second construction implies that *Susan* has received the letter. Chapter 5 gives a detailed analysis of this assumption and exploits it in the delineation of the prefixally-negated pairs of words in current usage.

Let us apply this to our topic. The negative prefixes *ab-* and *sub-* are tacked on to various bases to connote negation. They evoke the domain of *degradation*, an area of knowledge which describes a decline to a lower quality or degree. Yet, each prefix focuses on a specific facet of it. *Ab-* means 'move away from the state expressed by the base'. It targets degradation in quality. By contrast, *sub-* means 'inferior to the thing expressed by the base'. It targets degradation in degree. The difference in meaning can be spelled out by a negative pair. The adjectives *abnormal* and *subnormal* are derived from the adjectival base *normal*, which means 'ordinary or usual'. Despite the similarity in derivation, the two adjectives differ in terms of the construal imposed on their common base. In *She is an abnormal child*, the adjective *abnormal* means 'deviating from what is usual or typical, especially of behaviour'. *An abnormal child* is one whose behaviour is not normal. In *She is a subnormal child*, the adjective *subnormal* means 'lower than the average or expected standard, especially of intelligence'. *A subnormal child* is one who is mentally inferior.

Figure 3 presents a graphical representation which captures the two ways of construing the conceptual content *normal*.

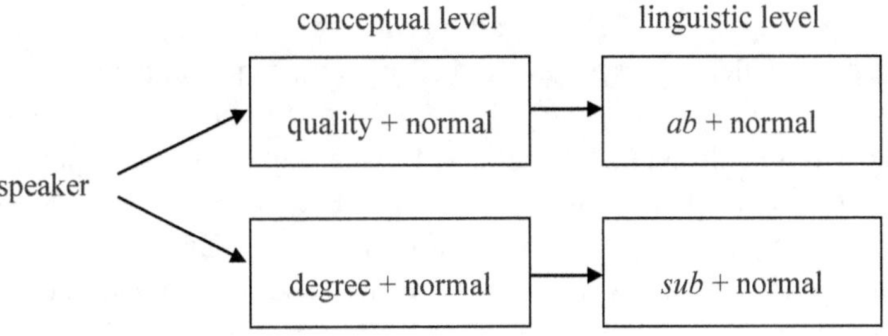

Figure 3: The construal of the conceptual content *normal*

1.4.2 Empirical method

Empirically, the study of prefixal negation is rooted in Usage-based Semantics, exemplified by linguists like Langacker (1988b, 2000), Kemmer & Barlow (2000), Tomasello (2000), Bybee & Hopper (20001) and Croft & Cruse (2004). Usage-based Semantics hinges on some pivotal premises. One premise is that linguistic structure emerges from language use. The linguistic units are extracted from patterns in the usage events experienced by a speaker. These units eventually build up the inventory that represents the speaker's linguistic system, i.e. his or her grammar. Another premise is that human language can be meaningfully accounted for by emphasising the interactive nature of language use. The linguistic units are recognised by, and accessible to, others in the linguistic community. A further premise is that the linguistic system encompasses both symbolic units and cognitive processes. The units are form-meaning pairings, variously known as assemblies or constructions, which constitute the inventory of a particular language. The processes are human abilities which relate and integrate these units in various ways. The following paragraphs give summaries of three central assumptions of Usage-based Semantics which apply to the area of prefixal negation.[7]

Knowledge of language is derived from language use. Usage-based Semantics views the mental grammar of the speaker, i.e. his or her knowledge of language, as a system derived from and grounded in *utterances*, actual instances of language use. Such instances which represent situated usage events are the basis on which the speaker's linguistic system is formed. A *usage event* is an utterance used in communication. It is a symbolic expression employed by a speaker in a particular circumstance for a particular purpose.[8] Usage events have a double import. First, usage events shape the linguistic system in the sense that the linguistic system is built up from such actual instances. General patterns, often called *schemas*, are gradually abstracted from the repetition of such instances of use. A *schema* is a cognitive representation with a general meaning, whose full details are reflected in its specific instances of usage. The instances elaborate the schema in contrasting ways. Second, usage events sanction the linguistic system in the sense that the schemas which arise via repeated activation of the general patterns are used to coin and understand novel expressions. Through repeated uses, these novel expressions become entrenched, i.e. firmly established, in the lexicon. From this assumption, it follows that units of language are dynamic, subject to creative extension.

In the present study, the notion of *utterance* receives central importance. An utterance has a unit-like status or is a somewhat discrete entity in that it represents the expression of a coherent idea making use of the conventions of the language. In terms of structure, an utterance may consist of a single word like *disbelief*, a phrase like *utter disbelief*, or a sentence like *She shook her head in disbelief*. All prefixally-negated words are considered utterances representing instances of linguistic behaviour on the part of the language user. They represent actual occurrences produced by the language user in communicative interaction. In speaking or writing, a language user, as a member of a particular linguistic community, attempts to achieve a particular interactional goal using particular linguistic strategies. The interactional goals include establishing interpersonal rapport, providing or eliciting information, and so on. The linguistic strategies include using speech acts, choosing words or grammatical constructions, and so on. The speaker's knowledge of the negative prefixes is based on situated instances of use.

The linguistic system is shaped by actual data. Usage-based Semantics takes the language people actually produce and understand as the primary object of study. Since the linguistic system is closely tied to, and directly derived from, usage events, language theories should be formulated on observing data from actual language production. The context in which an utterance is used is the best evidence available in accounting for its interpretation. It is on actual contexts that the linguist relies in making judgements, not on examples that lack natural contexts of production. Although speaker intuitions about constructed data are invaluable, they cannot be treated as the sole, or even primary, source of evidence in describing linguistic structure. The use of actual data has an important consequence. It helps to indicate what speakers do in their natural use of language and how ordinary comprehension and production of language work. Likewise, the use of context has an important consequence. It helps to account for word meaning. A word has not just a *conventional* meaning, referred to as a *coded* meaning, but also a *contextual* meaning, referred to as a *pragmatic* meaning. From this assumption, it becomes clear that language structure can be studied by taking into account the nature of language use and the fact that word meaning is changeable.

In the present study, the notion of *context* receives central importance. The context in which a word is used has important effects on its meaning. One kind of context relates to the other elements in the utterance. For example, in *declassify*, the negative prefix *de-* combines with a verb to mean 'reversal', whereas in *defrost* it combines with a noun to mean 'removal'. The examples demonstrate that the meaning of a negative prefix is not fixed, but changes relative to the element to which it is attached. Another kind of context relates to the background knowledge against which an utterance is produced and understood. For example, the sentence *He uncovered the thing* has two interpretations. When uttered by an official, it is likely to mean 'discover something secret', as in *He uncovered the plot*. When uttered by a mother, it is likely to mean 'remove the cover from something', as in *She uncovered the pan*. The examples demonstrate that context interacts with the speaker's intention and plays a crucial role in how the utterance is interpreted by the hearer. The analysis of prefixal negation is conducted on authentic data or natural contexts driven from the corpus and/or Internet pages.

Structure and substance are tightly linked. Usage-based Semantics considers the structure and substance of language as closely tied. The structure of language refers to the phonological form, whereas the substance of language refers to the semantic content. Both are related via language use, which is in turn influenced by experience. A linguistic theory should deal with the linkage between structure and substance since the intent of the speaker in employing a particular form is to convey a particular meaning. Any linguistic unit, lexical or grammatical, is meaningful and its structure reflects its substance. Therefore, the linguistic alternatives available to the speaker are not on an equal footing. They are motivated by semantic considerations, which include clues from both linguistic and non-linguistic worlds. These linguistic units achieve the status of entrenched units via *frequency* and *collocation*. Linguistic units that recur frequently become more entrenched in the linguistic system. Two types of frequency exist: *token* and *type*. *Token* frequency refers to instances, whereas *type* frequency refers to schemas. Similarly, linguistic units that co-occur become more established in the linguistic system. From this assumption, it can be inferred that the specific form of an expression is inseparable from its semantic organisation.

In the present study, the notion of *pairing* receives central importance. As linguistic units, all prefixally-negated words consist of two parts, form and meaning, which are conventionally associated. The form of a negative word is its orthographic representation. The meaning of a negative word is the idea conventionally associated with it. The linguistic unit might be a meaningful subpart of a word like *in-*, whole word like *indirect*, or strings of words like *indirect means*. The form of a complex word is paired with its own meaning. For example, when the base *moral* takes *a-*, the combination means 'without morals', as in *He was amoral*. When the base *moral* takes *im-*, the combination means 'morally wrong', as in *It's immoral to steal*. Each prefixally-negated word is associated with a special meaning. The form with *a-* is neutral and non-gradable, whereas the form with *im-* conveys criticism and gradability. As for entrenchment, token frequency is represented by the instances of a sense of a negative prefix, whereas type frequency is represented by the overall senses of a negative prefix.

1.5 Summary

This chapter has dealt with the semantics of rivalry in negative prefixation in English, a subject which has received hardly any attention in the literature. The first section set out the research, touching upon the theme, context and questions. The second section gave an overview of previous research on prefixal negation, which was carried out by lexicographers, grammarians and morphologists. The third section introduced the new semantic system, which is capable of confronting the questions raised. The new system is based on three semantic axes in terms of which prefixal negation will be described: *category*, *domain* and *construal*. The main goal was to show that there is mapping between semantic structure and morphological structure. The secondary goal was to show that the prefixes are responsible for the meaning differences between word pairs. The following chapters will turn to each of the three axes in my semantic system. The fourth section introduced the methodology. Theoretically, the study draws on the assumptions of Cognitive Semantics. Empirically, the study draws on the premises of Usage-based Semantics. The goal was to present a unified method that is best equipped to explain the interface between semantics and morphology in the analysis of negative prefixes.

To conclude the chapter, here are the main points which are crucial to the meaning of a linguistic expression:

1) A linguistic expression is a symbolic assembly. It has two poles: a semantic pole, its meaning, and a phonological pole, its form. The semantics associated with a symbolic expression is linked to a particular mental experience termed a concept, which derives in turn from a percept. The phonology associated with a symbolic expression is its perceptible form, which can be spoken or written. The use of a

symbolic expression then involves mapping its meaning to its form. For example, a linguistic expression like *dislike* is regarded as an association between the semantic structure 'not like' and the phonological structure /di'slaik/. To express a concept like this, language relates the two structures of the expression.

2) A linguistic expression is characterised by the imposition of a profile on a base. The base of a linguistic expression is the ground needed to establish the identity of the intended profile. The profile is the figure from which the composite expression inherits its character. Both the base and the profile are crucial to the meaning of a linguistic expression, but the profile is much more prominent. For example, in *dislike* the component *like* functions as the base for the component *dis-* which determines the character of the derivative by imposing the profile of distinction on it. Accordingly, every linguistic expression must specify, in the course of derivation, an entity that functions as the profile determining its semantic pole.

3) A linguistic expression, due to polysemy, displays a network of interrelated senses. Each node in the network represents one established sense. Each arc connecting two nodes indicates the nature of their association of which there are two types: schematicity marked by solid arrows and extension marked by dashed arrows. A is schematic for B, while B is an elaboration of A. The nodes vary greatly in their degree of salience. The precise form of the network may vary across speakers, depending on their experiences and judgements. This, however, does not pose a problem for communication provided that enough nodes are shared. A speaker's knowledge of a lexical expression therefore embraces the entire network.

4) A linguistic expression is characterised relative to the cognitive domain it belongs to. A *domain* is a coherent area of conceptualisation which provides the basis for the characterisation of a lexical expression. For example, in *disband* the negative prefix *dis-* presupposes the domain of *reversal*, in *disarm* it activates the domain of *removal*, and in *discomfort* it evokes the domain of *privation*. In each example, the negative prefix *dis-* incorporates a certain body of knowledge as a necessary part for characterising the resulting derivative. Accordingly, the speaker's use or detection of a lexical expression in the speech stream leads him or her to activate selected portions of the knowledge backgrounds they already possess.

5) A linguistic expression reflects a particular construal imposed on its content. *Construal* refers to the mental ability of a speaker to construe a conceived situation in many alternate ways. Two linguistic expressions may have the same conceptual content, but still contrast semantically. The semantic contrast between them is attributable to the imposition of alternate construals on their content. Each linguistic expression selects a different substructure as the one it designates. For example, both *non-natural* and *unnatural* describe the same state of affairs, that of distinction, but they are construed in different ways. *Non-natural* is neutral: something exists outside of nature. *Unnatural* is evaluative: something is abnormal or artificial.

Notes

1 Research into negation covers two main areas: affixal and non-affixal. The current study is concerned with affixal negation; therefore, non-affixal negation lies outside its scope. For the differences between affixal and non-affixal negation, the reader is referred to Welte (1978) and Tottie (1980).

2 In English, affixal negation is also expressed by suffixes. Both *-less*, and *-free* denote *privation*, but with a difference in meaning. The suffix *-less* indicates that the absence of the thing is of permanent nature, as in *a childless couple*. The suffix *-free*, by contrast, indicates that it is of temporary nature, as in *a child-free holiday*. For a detailed coverage of these and other suffixes in English, see Hamawand (2007).

3 In the present study, I make no attempt to describe prefixes in historical terms. Readers looking for a diachronic description of English negative prefixes are referred to Marchand (1969).

4 Given the general lack of attention to affixal semantics, there has been little work done on the semantics of alternation in word formation. There is, however, a substantial literature concerning other sorts of word formation. A large number of works such as Hockett (1954), Adams (1973), Matthews (1974), Aronoff (1976), Dressler (1985), Jensen (1990), Spencer (1991), Katamba (1993), Beard (1995), Haspelmath (2002), Booij (2005) and Aronoff & Fudeman (2005) serve as introductions to the basic notions used in morphology. A considerable number of studies such as Aronoff (1980), Cutler (1980), Fabb (1988), Baayan & Lieber (1991), Plag (1999) and Bauer (2001) deal with the productivity of English affixes conducting corpus-based analyses. Within word-formation, some handbooks such as Marchand (1969), Urdang (1982), Bauer (1983), Szymanek (1998), Adams (2001), Plag (2003) and Lieber (2005) shed light on aspects of derivation and offer detailed surveys of English affixes. A few sources such as Beard & Szymanek (1988), Carstairs-McCarthy (1992) and Stekauer (2000) survey the history of research done so far in morphology.

5 The same can be said of thesauruses. If one looks at *Longman Language Activator* (2002: 261), one will find that it describes rival pairs like *maltreatment* vs. *mistreatment* as synonymous.

6 The book-length treatments in Langacker (1987, 1991) present the theory and applications of Cognitive Grammar in more comprehensive detail.

7 Usage-based Semantics proposes a number of methodologies like surveys, experiments and corpora. As explained in Hamawand (2007), a *survey* is a way of getting information through a questionnaire from a sample of informants in order to verify a research question. An *experiment* is a way of collecting information through a test on a sample of respondents so as to validate a research question. Because the informants are aware of the procedure and the information which is elicited, there is a danger that the results of surveys and experiments only partially reflect actuality. For a comprehensive coverage of such methodologies, see Tummers et al. (2005).

8 As defined by Langacker (2000: 9), a *usage event* is 'the pairing of a vocalisation, in all its specificity, with a conceptualisation representing its full contextual understanding', or 'the pairing of a rich context-dependent conceptualisation with an actual vocalisation in all its phonetic detail'.

2 Derivation

2.0 Overview

This chapter gives a broad outline of the sub-discipline in morphology that studies word formation, called *derivation*. To do this, I adopt two tenets of Cognitive Semantics. One tenet is that a lexical expression is described in terms of three kinds of structure: a phonological structure (form), a semantic structure (meaning), and a symbolic structure (linkage). Applying this tenet to morphology, I argue that a complex negative word is a symbolic unit having two structures: the semantic structure is a reflection of its phonological structure. Another tenet is that the substructures of a lexical expression are integrated in terms of *valence* relations. Applying this tenet to morphology, I argue that the meaning of a complex negative word is conditioned by phonological and semantic compatibility between its subparts. To that end, the chapter is organised as follows. Section 1 sheds light on the fundamentals of morphology. It defines the subject of morphology and dwells on the terminology used in relation to it. Section 2 identifies its main tasks. One task is the combination of component parts of complex words. Here, two conceptions in morphology are presented: building-block and scaffolding. The other task relates to the interpretation of complex words. Here, two principles of interpretation are deliberated: compositionality and analysability. Section 5 gives a summary of the chapter and draws a number of conclusions.

2.1 Fundamentals

The word *morphology* consists of two parts: *morph* referring to form and *logy* referring to study. As a sub-discipline of linguistics, *morphology* is the study of the internal structure of composite words, which consist of basic elements called *morphemes*. Morphemes are the smallest units which cannot be further decomposed grammatically, but may be decomposed phonologically or semantically. They bear meaning or serve a grammatical function. Morphemes can either be *free* or *bound*. A *free* morpheme can appear as an independent word, whereas a *bound* morpheme can only appear as part of another word. For example, the word *useful* consists of two morphemes: one is *use* which is free, the other is *-ful* which is bound. In forming words, two major processes take place. One is *inflection*, the study of variation in the form of a lexical item. Inflection assigns the item a grammatical property so that it can fit a given syntactic slot.

For example, in *books* the inflectional morpheme *–s* serves the grammatical function of plurality. It does not change the meaning or part of speech. The other is *derivation*, the study of word formation. Derivation assigns a lexical item a semantic property so that it can fulfil a given discourse function. For example, in *selfish* the derivational morpheme *-ish* changes the part of speech of the word into an adjective as well as affecting its meaning.

Within *derivation*, three processes take place: *affixation*, *compounding* and *conversion*. *Affixation* is the process of forming a word by adding a bound morpheme to a free morpheme or a poly-morphemic base, as in *disagreement*. *Compounding* is the process of forming a word by combining two or more free morphemes, as in *homework*. *Conversion* is the process of forming a word by changing its part of speech without using any affix, as in *clean* (adjective) and *clean* (verb). Of the three processes, the present study is concerned with *affixation*. Affixes are bound morphemes which never occur on their own; they must be joined to other morphemes to complete their meanings. *Affixation* comprises three modes: *prefixation, infixation* and *suffixation*. *Prefixation* is the process of forming a word by attaching a bound morpheme to the front of a free morpheme, as in *disagree*. *Infixation* is the process of forming a word by inserting a linking morpheme between the two elements of a compound, as in *Auge-n-arzt* 'eye doctor' in German. *Suffixation* is the process of forming a word by attaching a bound morpheme to the end of a free morpheme, as in *agreeable*. Of these three modes, the present study is confined to *prefixation*. Within *prefixation*, the present study takes negative prefixes as its object of investigation.

2.2 Tasks

A fundamental task for a theory of morphology is to shed light on two important areas of the lexicon. One area, called combination, involves the way the structural units are arranged. The other area, called interpretation, involves the way the complex units resulting from combination are interpreted. Elaborating on the processes of production and comprehension in word formation then is a great challenge for morphology. Performing this task requires us to answer the question of how composite words are represented in the mental lexicon. On the one hand, are composite words represented as wholes, with no interest in their components? Put another way, is the lexicon organised on a word basis? Is a word like *amoral* represented in the mental lexicon in a full word format? On the other hand, are words represented as morphemic elements, with a focus on their components? In other words, is the lexicon organised on a morphemic basis? Is a word like *amoral* represented in the mental lexicon in a morphologically decomposed format, consisting of a base to which a prefix is linked?

2.3 Combination

Combination is the first task on which a theory of morphology concentrates. It relates to the horizontal relationships between the components of a composite structure, and facilitates our understanding of the way in which a composite structure is formed. In this regard, two questions are posed. The first is: which unit has priority: the word or the parts? The second is: what parameters are responsible for combining the parts? Two morphological conceptions account for the combination of the component parts of a construction. One is the building-block conception, which is dominant in Formal Morphology. The other is the scaffolding conception, which is prevalent in Cognitive Morphology.

2.3.1 The building-block conception

In Formal Linguistics, *morphology*, as Bauer (2003: 335) states, is the study of the forms of words. It is the study of the ways in which words are built up from smaller units. The main aim is to identify the various component parts of a word. One part is called the *root*, the basic part of a lexeme which cannot be further analysed into smaller parts. This is the part which remains when all affixes are removed. For example, *happy* is the root in *unhappy*. A *base* is a morpheme consisting of a root, a stem or a root plus a morpheme, which serves, upon the addition of a further morpheme, to form another word. For example, *unhappy* is the base in *unhappiness*. The other part is the *affix*, a bound morpheme which cannot function separately but must be added to a free morpheme. To study combination, one important analysis in Formal Morphology pertains to the building-block conception. In this conception, as Langacker (1987: 452–457) explains, a composite word is constructed out of its components simply by stacking them together in some appropriate fashion. The essence of this conception can be summarised in three defining characteristics.

First, morphology is the study of how composite words can be segmented into discrete components, called building blocks. For example, *un-* and *ease* are the building blocks of the composite word *unease*. A composite word is then seen as made up of separate components, which are combined according to certain rules. The gist of the building-block conception of morphology is mirrored by Jensen's (1990: 21) principle that a word must be exclusively divided into morphemes. On this basis, a morphological analysis of a composite word can only be achieved by identifying its components. In this conception of morphology, priority is given to the component parts of a word, which must be arranged neatly. When there is no segmentation, there is no morphological relationship at all. It is an either-or matter.

Second, the construction of a composite word and the means to assemble its components is carried out independently of the language user. Building a word out of blocks is analogous to building a house out of bricks. Morphology is seen as a matter of objective composition. The components, as Lakoff & Johnson (1980: 204) explain, are looked upon like objects, which have properties in and of themselves and stand in fixed relationships to one another independently of the speaker or writer who uses them. As objects, words have parts. A morphological analysis then deals with how a word is constructed from various components. It does not reside in how the language user conceives a situation and employs the word in question. This is so because grammar operates independently of cognitive principles.

Third, the task of the formal linguist is to study the building-block structure of composite words, the inherent properties of their components and the relationships among them. Precisely, the task is to identify the morphemes and pin down the devices which condition their ability to combine. A morphological analysis consists of listing building blocks and stating how they are combined. According to Halle (1973: 3), speakers know the words of their language, recognise the components of the words and realise the ways they are arranged. Therefore, the grammar of a language must explain how a composite word is made up of components and present the rules that combine them. Analysing a composite word amounts to listing its components and stating how they are combined.

2.3.1.1 *Representative works*

In the literature, the building-block conception is represented by two morphological approaches. One approach views derivation simply as an operation. Another approach views derivation simply as a lexical selection. For a thorough treatment of the approaches, see Hockett (1954) and Spencer (1991), among others.

Item-and-Process approach. In the Item-and-Process approach, which is alternatively called Lexeme-based morphology, the underlying form of a word is the basic element of morphological description. A word is formed by deriving its surface form from an underlying base by means of rules. Each rule changes the form of the base and concomitantly has some characteristic semantic or morpho-syntactic effect. The form-meaning correspondence is presumed to be one-to-one. The form of a word is analysed as being the result of applying rules to its deep structure to transform it into its surface structure. A derivational rule takes a base, makes some changes to it and outputs a new form. For example, *careless* is the result of the base *care* and a transformation that changes it into the surface form *careless*. This approach does not have any difficulty when faced with irregular forms. While the plural of *dog* is formed by adding an *-s* to the end, the plural of *geese* is formed by changing the vowel in the base. This approach was dominant in the generative tradition. Works that fall into the Item-and-Process camp include, among others, Aronoff (1976), Anderson (1992) and Beard (1995).

Morphologists who adhere to this approach trace back morphological derivations to the same deep structure and attribute the surface differences to transformational rules. Since transformations do not change meaning, the resulting variations are similar. For example, Lees (1960) considers morphological constructions as the output of phrase structure rules operating on lexical items. The meaning of a given construction is determined by its deep structure; so transformationally-related constructions, i.e. those sharing the same deep structures, are semantically equivalent. Relating this to prefixation, the members of a pair are supposed to have one deep structure, and hence be semantically alike. The surface differences are the result of different transformational operations. Alternative prefixes are treated as interchangeable, and the choice between them is the result of different syntactic transformations. The presence of alternative prefixes is a matter of idiosyncrasy and more or less an instance of synonymy.

Item-and-Arrangement approach. In the Item-and-Arrangement approach, which is alternatively called Morpheme-based morphology, the morpheme is the basic element of morphological description. A word is built up by inserting morphemes into appropriate positions within it. Each morpheme contributes a distinct meaning to the complex word. The form-meaning correspondence is presumed to be one-to-one. The form of a word is analysed as consisting of a set of morphemes arranged in sequence by rules. A word like *carelessness* is made up of the morphemes *care*, *-less* and *-ness*, which are placed after each other like beads on a string. *Care* is the base and the other two morphemes are derivational suffixes. Although this sort of analysis applies to regular cases like *dogs*, it fails to account for irregular cases like *geese*, for example. Adherents of this type of morphology defend their approach by saying that *geese* is *goose* followed by a null morpheme, a morpheme that has no phonological structure. This approach was pursued in American Structuralism. Works that fall into the Item-and-Arrangement camp include, among others, Williams (1981), Selkirk (1982) and Lieber (1992).

Morphologists who adhere to this approach exclude transformational rules and deep structures from derivational morphology. Morphological constructions are not related by transformational rules. Word-formation cannot be governed by purely syntactic transformations; rather it should be governed by specific considerations which mediate the relationship between the base and the affix. They view morphological constructions as the outcome of morphological, phonological or semantic considerations. Each consideration has a different impact with respect to the status of the affix or the value of the resultant derivative. Relating this to prefixation, the members of a pair are hypothesised to be more or less different. Alternative prefixes are treated as substitutable, and the choice between them is explained in terms of three types of selection hypotheses: the morphological shape of the base, the phonological property of the affix, and the semantic category of the base.[1]

2.3.1.2 Combination restrictions

The question then raised is: what kinds of restrictions determine the ways in which the parts are combined together to form the whole? In the formal literature, theories on word formation are in favour of rules. A new word is formed by applying a regular rule to an already existing word. Aronoff (1976) characterises processes of word-formation in terms of rules. In his opinion, a theory of morphology must not include the premise that morphemes are necessarily meaningful.[2] Bauer (1983) equates processes of word-formation with syntactic processes of sentence generation. Both morphologists propose to add a word-formation component to the lexicon of generative grammar. Their goal is to posit a unified theory of morphology that is capable of dealing with two distinct but related matters. One is the analysis of the existing composite words. The other is the formation of the new composite words. Due to some aspect of their make-up, bases differ in providing a suitable input to a rule of word-formation. The rules which determine the combination process mostly relate to morphology and phonology, and partially to semantics.

Morphological. In the light of this restriction, the use of a negative prefix is determined by the morphological form of the base. According to Huddleston & Pullum (2002: 1688), the negative prefix *a-* is from Greek and so is attached largely to Latinate bases, as in *asymmetrical*. The negative prefix *in-* takes Latinate bases, whereas the negative prefix *un-* takes English bases as shown by the distinction in *inedible* vs. *uneatable*. For reasons of euphony, the negative prefix *in-*, as Marchand (1969: 169) writes, cannot be used with bases beginning with *in-*, as in **ininflammable*, **inintelligible*, **ininflected*, etc. Because it is restricted to Latinate bases, the negative prefix *in-*, Bauer (2003: 219) contends, does not combine with words beginning with the letters *th*, *sh*, or *w*. In expressing reversal, negative prefixes are attached to verbs, but of different nature. As cited in Huddleston & Pullum (2002: 1689), the negative prefix *de-* is used productively with verbal bases ending in *-ate, -ify* and especially *-ise*, as in *desegregate, declassify* and *decentralise*. The negative prefix *dis-*, by contrast, is generally selected before verbal bases beginning with *-en*, as in *disengage*.

Phonological. By virtue of this restriction, the use of a negative prefix is influenced by the phonological shape of the base. As shown by Marchand (1969: 169), negative prefixes generally do not change their forms. The only exception is the prefix *in-* which assimilates to the initial consonants of its bases and changes into *im-* before labials, such as *immodest*; *ir-* before *r*, such as *irrelevant*; and *il-* before *l*, such as *illegal*. Elsewhere, it remains *in-*, such as *indefinite, inoperable, intolerable*, etc. A kind of optional assimilation, as explained by Huddleston & Pullum (2002: 1687), is also found with the negative prefix *un-*: compare /ʌnˊpopjulə/ and /ʌmˊpopjulə/ from *unpopular*. As demonstrated by Plag (2003: 100), negative prefixes do not influence the stress patterns of their bases. For example, the negative prefix *mis-* is usually either unstressed or secondarily stressed. In some words, however, the negative prefix *mis-* receives a primary stress. These words are either lexicalisations such as *ˊmischief*, or nouns that are homophonous with verbs such as *ˊmiscount* vs. *misˊcount*. Finally, negative prefixes do not change the phonological specifications of their bases. One exception is the negative prefix *dis-* which removes an already existing prefix in the formation of a negative word, as in *encourage* > *discourage*.

Semantic. In view of this restriction, the use of a negative prefix is conditioned by the semantic feature of the base. An example is provided by Zimmer (1964: 15). In his opinion, 'negative prefixes are not used (in English) with adjectival stems that have 'negative' value on evaluative scales such as 'good-bad'…'. In other words, negative prefixes in English tend not to attach to bases that are negative in content. For instance, it is possible to say *unwell*, but not **unill*, *unhappy* but not **unsad, uncheerful* but not **unsorrowful*, *unoptimistic* but not **unpessimistic*. Although this limitation is fairly strong, it is not hard to find, as Zimmer (1964: 30, 35–7) writes, a number of exceptions. Words like *unhostile, unvulgar*, and *incorrupt* seem to have negative content, yet they take negative prefixes. This can be seen as an instance of *blocking*, which refers, according to Aronoff (1976: 43), to the non-occurrence of a complex form because of the existence of another form. For instance, the existence of lexemes like *bad* and *small* blocks the formation of lexemes like **ungood* and **unbig*. In Aronoff's view, the reason is that there is simply no need for such new lexemes as the existing ones denote the concepts in the language. The present study is in discord with Aronoff's view. In English, there exist word pairs which share same bases, yet they serve different purposes in the discourse.

2.3.2 The scaffolding conception

In Cognitive Linguistics, *morphology* is, I argue, the study of the meanings of words. It is the study of the relationships between the components of a composite word. The main aim is to show how the meanings of two or more components integrate to form a composite word. For this reason, the component parts are defined differently. In Langacker's (1987: 345) terminology, a *root* is a phonologically and semantically autonomous unit within a composite structure. It is autonomous in the sense that it can stand on its own. It is the smallest unit because it is not analysable into smaller symbolic components. For example, in *inhuman*, *human* is the root to which *in-* is attached. A *base* is an autonomous phonological structure at any level within a word. For example, in *anti-abortion*, *abortion* is the base. An *affix* is a phonologically and semantically dependent unit within a composite structure. It is dependent in the sense that it needs the root or the base to complete its meaning. To study combination, one important analysis in Cognitive Morphology pertains to the scaffolding conception. Langacker (1987: 461) defines scaffolding as follows: 'component structures are seen as scaffolding erected for the construction of a complex expression; once the complex structure is in place (established as a unit), the scaffolding is no longer essential and is eventually discarded'. The essence of this conception can be summed up by three defining characteristics.

First, morphology does not view a composite word as an edifice constructed out of smaller components. The component parts are not the building blocks out of which a composite word is assembled, but function instead to motivate selected facets of its meaning. The use of a composite word involves two phases of activation, Langacker (1987: 452) contends. First, the word as a whole is activated. It has its own semantic value. Second, the component parts are activated. They reflect certain aspects of the composite value. For example, the composition of the word *counter-attack* is rather seen as a case of primary and secondary activation. Primarily, it is activated as a unit. It denotes response. Secondarily, its component parts *counter* – and *attack* are activated. The prefix highlights opposition, while the base highlights action. In this conception, the composite structure has, as Langacker (1997: 11) asserts, priority over its component structures.

Second, the construction of a composite word is produced by the language user, based on general patterns of usage. The component parts of a composite word are mere scaffolding for the job of word construction. Morphology is seen as a matter of conceptual composition. It is a network in the mental lexicon, where the component parts are lexically interrelated. In Bybee's (1985) view, when a composite word is used, its morphological units are activated by the speaker or writer through lexical connections.

A morphological analysis should then treat a composite word as a coherent structure in its own right. Its meaning is not solely derivable from the meanings of its components. It is characterised relative to a given cognitive domain distinct from any that its component structures have in isolation. A morphological analysis resides in how the language user matches a given word with a given situation.

Third, the task of the cognitive linguist is to pin down the principles that allow the speaker to associate the linguistic elements and enable him/her to choose the appropriate morpheme to match with a given situation. Precisely, the task is to identify the contributions made by the component parts to the overall meaning of a composite word. According to Hamawand (2007, 2008), the morphology of language should concentrate on three pivotal areas in the processes of derivation: *categorisation*, *configuration* and *conceptualisation*. *Categorisation* refers to the cognitive ability to group together the various senses of a given affix into one category. *Configuration* refers to the cognitive ability to assemble the various affixes into domains, background knowledge with respect to which they can be characterised. *Conceptualisation* refers to the cognitive ability to conceive or experience a situation and choose the appropriate morphological structures to express it.

2.3.2.1 Representative work

In the literature on Cognitive Linguistics, not much has been written on the topic of morphology. This is due to the fact that syntax has been the centre of attention of most cognitive linguists. In spite of that, some morphological studies can be identified. One study is Bybee (1985), which reports on a cross-linguistic survey of verbal morphology in several languages. Another study is Ryder (1994), which deals exclusively with noun compounds in English. A further study is Tuggy (2005), which provides a general view of word formation. The approach adopted in the present study views derivation simply as a relationship between the conceptual content of a base and the construal imposed on it. This approach, which is proposed in Hamawand (2007, 2008), deals exclusively with English morphology.

Lexical connections approach. One representation of the scaffolding conception is the lexical relations approach proposed by Bybee (1985). In this approach, which she exemplifies with inflectional categories, Bybee argues for a connectionist theory of morphology, in which lexical rules have no status independent of the lexical items to which they are applicable. Lexical rules represent schemas, abstractions from memorised lexical items which have semantic or phonological properties. An inflectional rule is simply a relationship which is firmly established in long-term memory. It can be reduced to the arrangement of memorised items in mental storage. In cases of morphological creativity, speakers use the cognitive process of analogy, by searching their memories for relations which are similar to the intended coinage. To facilitate co-occurrence, Bybee proposes two determinants. One determinant pertains to *relevance*, where the semantic content of an affix directly affects or modifies the semantic content of a stem. Another determinant pertains to *generality*, where an affix must be applicable to all stems of the appropriate category and must obligatorily occur in the appropriate syntactic context. Although the processes proposed in Bybee's approach are vital in inflection, her approach cannot, unfortunately, account for derivational data which include word pairs that share the same bases.

Category-Domain-Construal approach. Another representation of the scaffolding conception is the *Category-Domain-Construal* model, which I have advanced in Hamawand (2007, 2008). In this model, the prime objective is to look in detail at how morphology and semantics interact. This approach aims to show the direct relevance of meaning to the phenomenon of derivational morphology. It is a model in which the morphological structure mirrors the semantic structure of a complex word. The gist of the model is to show that when a speaker perceives an entity in the environment, s/he passes the information about it to the mind. The mind then processes the information by assigning it to one of the mental acts. The model develops a practical framework for analysing the morphological structure of negative prefixes by building on three theories of meaning: *category*, *domain* and *construal*. *Category* exhibits the multiple senses of negative prefixes. *Domain* displays the distinct uses of the negative prefixes. *Construal* shows how the negative prefixes represent discrete conceptualisations.

The first theory, *categorisation*, pertains to the description of individual items. It refers to the mental act of grouping together the various senses of a given lexical item into one category. A category is defined by a *prototype*, rather than a set of strictly necessary conditions. A *prototype* is the most typical member of a category, the one which comes to mind first and possesses all the properties of the category. Category membership is a question of similarity to the prototype. It is a matter of degree, rather than all or nothing. The members resemble the prototype by having only some, not all, of the properties. A category then has a graded structure. The members have different, unequal, status. Some enjoy full status; others enjoy marginal status. A category has fuzzy boundaries, rather than clear ones. Although some members appear less similar to the prototype, they are included in the category because of family resemblance. The adoption of prototype theory has a crucial consequence for language description. This is seen in the tendency of word meanings to be variable and flexible.

In Cognitive Grammar, categorisation is understood in terms of schemas. A *schema* is a coarse-grained pattern of a concept, which contains specifications common to its elaborations. The elaborate structures may differ from one another in finer detail. For example, the concept [In+N] is a schema whose specifications are compatible with those of its elaborations such as *illegal, incredible* and *inattentive*. These are alike in elaborating the same schema, but they differ from each other in certain ways. The schema-elaboration relation is represented in Cognitive Grammar by nodes and arcs. Each node in the network represents one established sense of the lexical item, and each arc connecting two nodes indicates the nature of their association. Nodes are associated via categorising relationships, of which there are two basic types: schematicity (indicated by solid arrows) and extension (marked by dashed arrows). Categorising relationships vary according to three dimensions: the degree of entrenchment, the difference in value, and the extent of distance from the basic sense.

Let us take an example to demonstrate the categorial character of negative prefixes. Prototypically, the prefix *semi-*, from Latin, is tacked on to adjectives to form new adjectives, acquiring the meaning 'half the thing described in the adjectival base'. For example, *semi-circular* means having the character of one half of a circle. The same meaning applies to a nominal base. For example, *semi-tone* means half a tone. Extended from this is the sense 'occurring twice during the period described in the adjectival base'. For example, *semi-weekly* means occurring twice a week. Peripherally, the prefix *semi-* is tacked on to adjectives to form new adjectives, acquiring the meaning 'almost but not entirely the thing described in the adjectival base'. For example, *semi-conscious* means almost but not entirely conscious. The same meaning is true of a nominal base. For example, *semi-final* means the stage before the final. Extended from this is the sense 'just like the thing described in the adjectival base'. For example, *semi-tropical* means like the tropics. The same meaning arises when the base is nominal. For example, *semi-vowel* means like a vowel, something in between a consonant and a vowel in character.

Figure 4 presents a graphical representation which captures the multiple senses of the negative prefix *semi-*. Note that the solid arrow represents the prototypical sense, whereas the dashed arrows represent the semantic extensions.

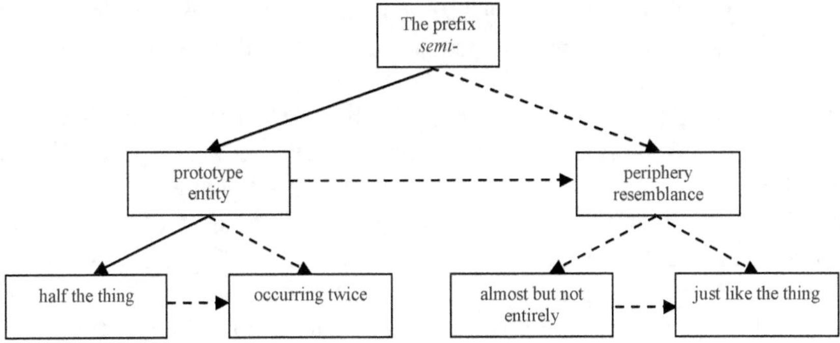

Figure 4: The semantic network of the negative prefix *semi-*

From the preceding discussion, one concludes that a prefix forms a category of its own, which includes its multiple senses. The senses gather around one representative sense, referred to as the *prototype*. A look at the categorial descriptions of the negative prefixes shows that some of them denote more or less the same meaning. In spite of this, the prefixes are not, according to cognitive premises, identical in meaning. The question then is: how can one show the difference in meaning between them? The answer lies in the *domain* theory, which helps to identify their nuances. It is within these domains that the prefixes can stand against each other as rivals. So, a domain is concerned with a knowledge configuration in which prefixes gather showing similarity on the surface but dissimilarity below the surface. Two negative prefixes may stand for one concept but differ in the specifics. The elaboration of this theory will be the subject matter of the following discussion.

The second theory, *configuration*, relates to the description of collective items. It refers to the mental act of clustering together various lexical items to form domains. A *domain* is background knowledge with respect to which lexical items can be characterised. Basic to the theory is the view that to understand the meaning of a lexical item one has to understand the domain in which it is embedded. A domain has various facets of meaning, each of which is represented by a specific lexical item. The meanings of the lexical items are not related to each other directly. Rather, they are related via their links to the different facets of a given domain, which represent different concepts. The adoption of a domain theory has a crucial consequence for describing language. It is very useful in the contrastive analysis of different lexical items. This is seen by the fact that different meanings are associated with them.

In Cognitive Grammar, *configuration* both joins and splits lexical items. It joins lexical items in one domain to the degree that they elaborate the same concept. It splits lexical items into separate facets to the degree that they have discrete specifications, which are firmly established by usage. Classification is established in Cognitive Grammar not only by relationships of generality but also by relationships of specificity. For example, the structures A, B and C are assigned to different facets, yet they are members of one and the same domain. However, this does not mean there are no cases of overlapping between the structures. That is, although the structures A, B and C represent different cases of the domain X, they overlap on certain occasions. In Cognitive Grammar, differences along parameters along which linguistic structures vary are relative rather than absolute.

Let us consider the tendency of negative prefixes to gather in domains through an example. The prefixes *pseudo-*, *quasi-* and *semi-* realise, I argue, the domain of *inadequacy*, an area of knowledge which refers to something or someone as having nothing or only some of a thing described. In all the formations, the prefixes carry a negative shade of meaning. However, they differ in that each has a particular nuance. The prefix *pseudo-* is used mostly to describe something or someone as being deceptive, false, or a sham. *A pseudo-intellectual person* is a person who engages in false intellectualism or is intellectually dishonest. The prefix *quasi-* is used chiefly to describe something or someone as resembling the thing described. *A quasi-judicial role* is a role that is like that of a judge. The prefix *semi-* is used mainly to describe something or someone as having the thing construed but only to some degree. *Semi-literate people* are people who are partly but not completely literate.

Figure 5 presents a configuration which captures the different facets of the domain of *inadequacy*.

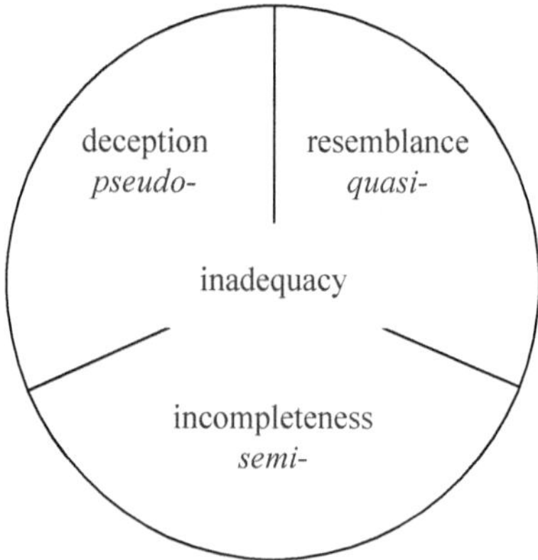

Figure 5: The domain of inadequacy

In the foregoing discussion, I showed how the *domain* theory applies to the description of prefixes in English. The description comprised three steps. First, I placed the prefixes in one domain, which I named *inadequacy*. Second, I assigned each prefix to a particular facet within the domain. This is done relative to the type of derivation they form and the behaviour they have in context. Third, I explained the rivalry between the prefixes by pinpointing the peculiarity of each prefix which makes it different from its counterparts. The question then is: how and when can one use a prefix? The answer resides in the *construal* theory. *Construal* concerns the ways a speaker conceptualises a given situation and the morphological expressions s/he chooses to realise them. Two negative prefixes that stand as rivals conceptualise a situation in different ways. The elaboration of this theory will be the task of the following discussion.

The third theory, *conceptualisation*, concerns the interpretation of the derived words. It refers to the mental act of conceiving or experiencing a situation and choosing the appropriate linguistic structure to express it. Fundamental to the theory is the assumption that a linguistic structure imposes a certain construal on a situation it encodes. Conceptually, there are countless ways of construing a given situation, in which each conception may deviate from another in any manner or to any degree. Linguistically, a variety of lexical devices are usually available as alternate means of encoding a given conception. Two expressions having the same conceptual content encode different meanings because they encode different construals of a situation. A situation's objective properties are consequently insufficient to predict the morpho-lexical structure of an expression describing it. A natural consequence of the construal theory is the idea that different words give rise to distinct construals. This is seen by the fact that each construal of a situation projects onto a particular form.

The choice of an alternative is argued to be a matter of two keystones. One keystone concerns *conceptual content*, the semantic property inherent in the components of an expression. The content of the base is multi-faceted, whereas the prefix has its own content which it imposes on the base and so gives the resulting alternative a different meaning in the derivational process. That is, when two alternative prefixes attach to a base, each serves to highlight a different facet of the base's content. Each of the resulting alternatives encodes, therefore, a distinct meaning. Another keystone concerns *construal*, the way the speaker describes the content vis-à-vis the communicative needs. The meaning of a derived alternative involves the particular construal the speaker employs to describe a situation. Linguistically, the construal is encoded by means of a prefix. That is, when two alternatives have the same conceptual content, they differ semantically by virtue of the construals they represent and the prefixes they host. The prefixes, therefore, are responsible for disengaging the lexical alternatives.

Let us look at an example to explicate the role of construal in clarifying meaning distinctions. The prefixes *quasi-* and *semi-* are attached to adjectival bases to form new adjectives. They evoke the domain of *inadequacy*, but each highlights a specific aspect of it. The prefix *quasi-* means 'resembling the thing named by the adjectival base', whereas the prefix *semi-* means 'having some of the thing described in the adjectival base'. The difference in meaning can be shown by an example of an adjectival pair. In *a quasi-religious language*, the adjective means resembling something that is religious. In *a semi-religious language*, the adjective means having a somewhat religious character. Likewise, in *a quasi-scientific system* the adjective means resembling something that is scientific, whereas in *a pseudo-scientific system* the adjective means pretending to have the thing so as to deceive people.

Figure 6 presents a graphical representation which captures the two ways of construing the conceptual content *religious*.

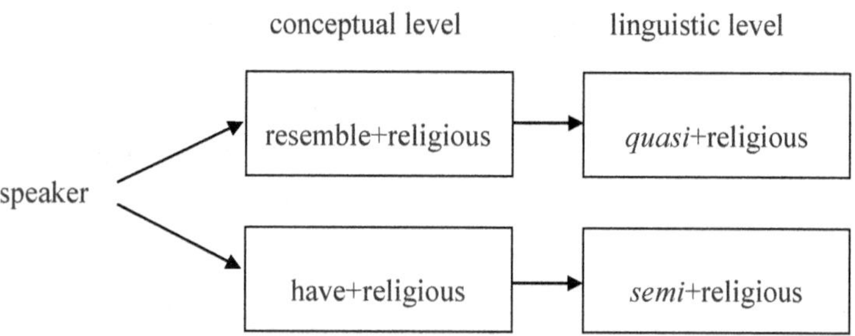

Figure 6: The construal of the conceptual content *religious*

2.3.2.2 Integration parameters

The question then raised is: what parameters affect the integration of the substructures of a composite word? Langacker (1987: 277–327) addresses the issue of integration in terms of the establishment of *valence* relations. *Valence* is the mechanism whereby two linguistic units combine to form a composite unit. The term *valence* is taken from chemistry, where it refers to the combining properties of atoms. Langacker (1988b: 102) writes: 'A valence relation between two predications is possible just in case these predications overlap, in the sense that some substructure within one corresponds to a substructure within the other and is construed as identical to it'. A central concern of Cognitive Morphology then is to pin down the parameters on the basis of which a valence relation can be established. The presence or absence of such parameters often has striking consequences for the semantic value and the linguistic behaviour of the composite word. In the cognitive model, the parameters have been applied to phrase-level constructions, precisely by looking at how words combine to make phrases and accounting for the relationships within the phrase. In the present study, the parameters are extended to word-level constructions, i.e. to the relationships between the subparts of words.

Correspondence. The parameter of *correspondence* relates to the idea that two substructures can be integrated to form a composite structure if they have certain elements in common at both semantic and phonological poles. A composite structure is formed by unifying identical substructures and merging their specifications. Typically, one substructure corresponds to, and serves to elaborate, a schematic substructure within the other. Let us apply this parameter to our example. The composite structure *insecure* is composed of two component substructures: *in-* and *secure*. The integration of the two substructures is affected by correspondences established between them. Phonologically, the prefix *in-* makes a schematic reference to a base, which *secure* elaborates to yield *insecure*. Semantically, the *in-* evokes the relation of *distinction* which corresponds to the relation profiled by *secure*. By unifying the two corresponding relations, one obtains the composite structure *insecure*. The *in-* component is relational in that it requires an entity which completes its meaning. The relation between the two can be understood in terms of a schematic trajector and landmark. *In-* elaborates the trajector, while *secure* elaborates the landmark.

Determinacy. The parameter of *determinacy* concerns the idea that two substructures can be integrated to form a composite structure if they show asymmetry in terms of *profile determinacy.* This concerns which of the two component substructures lends its profile to the entire composite structure. A composite structure usually inherits its profiling from one of the component substructures, which is called the *profile determinant*. It is an element in a construction which is central because it is primarily responsible for the character of the construction, and because the other elements are in a syntactic and semantic relationship with it. On applying this parameter to our example, we find that in *insecure* the prefix *in-* is the key substructure in that it lends its profile to the entire composite structure. The base *secure* is describable as one whose profile is overridden by that of the prefix. The composite structure denotes distinction and profiles a property. It acquires this meaning because the negative prefix *in-* is its profile determinant. If *insecure* occurs as a component structure of a yet more complex construction like *insecure building*, the profile determinant of the whole construction is the noun *building*, which profiles a thing.

In the formal model, the *profile determinant* is known as the *head*, which determines the core meaning as well as the grammatical category of the phrase it heads. Both models hold that the *profile determinant* or *head* lends its features to the phrase that contains it. The *head* of a phrase is a single word that determines the categorial status of the phrase. For example, a noun heads a noun phrase. In addition, the *head* determines the core meaning of the phrase, and selects its dependents, i.e. the elements it co-occurs with inside the phrase. The difference, of course, is that the features of the *profile determinant* are defined in semantic terms and relate to its schematic meaning, e.g. process or thing. By contrast, the features of the *head* are defined in structural terms and relate to its grammatical category, e.g. verb or noun.

Dependence. The parameter of *dependence* pertains to the idea that two substructures can be integrated to form a composite structure if they show asymmetry in terms of *autonomy* and *dependence*. Of the two component substructures, one qualifies as *autonomous* (A), the other as *dependent* (D). The asymmetry between A and D substructures resides in the fact that (A) exists on its own without need of (D) to complete its meaning, whereas (D) is dependent on (A) to complete its meaning to the extent that (A) constitutes a salient substructure within the semantic structure of (D) and adds conceptual substance to it. In the above example *insecure*, the base *secure* represents the (A) substructure. Phonologically, it can stand by itself as a fully autonomous form. Semantically, it can stand as a fully acceptable unit. It is possible to conceptualise it without making any necessary reference to anything outside the concept itself. By contrast, the prefix *in-* represents the (D) substructure. Phonologically, it cannot occur as an autonomous form. Semantically, it cannot stand by itself as a fully acceptable unit. It has to attach itself to a host of an appropriate kind. The prefix has a hole called an *elaboration site*, or more briefly an *e-site*, and the base fills that hole.

Two types of structure, i.e. words that can integrate with the *profile determinant*, exist. These are *complements* and *modifiers*. As defined by Langacker (1994: 592–3), a *complement* is an autonomous substructure that elaborates a salient part in the semantic structure of the *profile determinant*. A *complement* adds intrinsic conceptual substance to the *profile determinant*. In the expression *insecure*, the substructure *in-* is the *profile determinant*, whereas the substructure *secure* is the complement. In the head-complement conception, the *complement* is obligatory in the sense that it completes the meaning of the *profile determinant*. A *modifier*, by contrast, is a dependent substructure that has a salient part in its semantic structure which is elaborated by the *profile determinant*. A *modifier* adds non-intrinsic specifications to the *profile determinant*. In the phrase *insecure building*, for instance, the noun *building* functions as the *profile determinant*, whereas *insecure* functions as the modifier. In the head-modifier conception, the *modifier* is optional in the sense that it only provides additional information to the *profile determinant*.[3]

As in the cognitive model, in the formal model the relationship between the component parts of a phrase are described in terms of dependence. The difference is that the cognitive model describes the relationship in terms of conceptual dependence, whereas the formal model describes it in terms of categorial selection. In the formal model, *autonomy* and *dependence* are described in terms of heads and dependents. The term *dependent* has its roots in a selection-based theory. For example, the presence of a preposition entails the presence of a noun phrase; so the preposition is said to select or subcategorise for the noun phrase. This information is stored in the lexicon in 'selection frames'. In the formal model, a *head* is entirely autonomous within its phrase in the sense that it selects all its dependents, some obligatorily like complements and others optionally like modifiers. The cognitive model is rather different: a *head* can conceptually rely on its dependents if they elaborate some aspect of its structure. In the cognitive model, the *head* is dependent on the complement to elaborate an essential part within its structure, but the *modifier* is dependent on the head to elaborate some schematic aspect of its structure.

Constituency. The parameter of *constituency* refers to the order in which component substructures are successively integrated to yield composite structures. *Constituency* is the combination of smaller subparts into larger, more complex units. Though related to the traditional notion of *constituent structure*, constituency is not to be identified with an autonomous level of syntactic organisation. Rather, constituency has to do with the assembling of meaningful symbolic units into structures of progressively increasing semantic and phonological complexity. Cognitive processing involves multiple levels of organisation, such that elements at one level combine to form a composite structure that functions as a unitary entity at the next higher level and so on. A *constituent* is a composite structure formed by integrating two or more components. Let us return to the example to illustrate this parameter. In *insecure building*, at the first or lower level of constituency, the base *secure* is integrated with the negative prefix *in-* to form the adjective *insecure*. At the second or higher level, *insecure* combines with *building* deriving thus the composite structure of the overall expression.[4]

In the formal model, *constituency* represents a structural property of language. The arrangement of parts of a composite structure is regulated by 'words and rules', which are primitive. Two things are important here: the parts which are combined, and the order in which they should appear. In the cognitive model, *constituency* pertains to the order of combining or decomposing a whole into parts. The order is variable; different speakers may do it differently, or do it differently on different occasions. As long as it does not pose a barrier to communication, the order does not matter. What matters is how the parts are combined or analysed. For instance, the word *misconstruction* can undergo two types of analysis. In the first analysis, it is decomposed into *mis* and *construction* and the latter is decomposed into *construct* and *ion*, yielding the constituency mis[[construct]ion]. In the second analysis, it is decomposed into *misconstruct* and *ion*, and the former is decomposed into *mis* and *construct*, yielding the constituency [[mis [construct]ion].[5]

2.4 Interpretation

Interpretation is the second task on which a theory of morphology concentrates. This relates to the vertical relationships between a composite word and its parts, and facilitates our understanding of the substantial semantic content which a composite structure has. In this respect, two questions are posed. The first is: is it possible to predict the meaning of a composite structure from its component parts? The second is: is it easy to identify the contribution made by each component part to the meaning of the composite structure? Two morphological principles account for the interpretation of a composite structure. One is *compositionality*, the degree to which the component structures are exhaustible in the meaning of the composite structure. The other is *analysability*, the degree to which the roles of the component structures are discernible in the composite structure.

2.4.1 Compositionality

Compositionality concerns the relation between component structures and the composite structure that derives from them. As Langacker (1987: 448–452) states, it concerns the regularity of compositional relationships, i.e. the degree to which the value of the whole is predictable from the values of its parts. It refers to the degree of regularity in the assembly of a composite structure out of smaller components. This means that the contents of *a-* and *moral* are included in that of *amoral*, for instance. This type of representation is referred to as *immanence*, where the two component structures exist within the composite structure. This is the case in which only the occurrence of the component parts is necessary, with occurrence implying use outside a context. Compositionality exists in two types: full and partial.

The formal view assumes full compositionality. According to this assumption, the meaning of an expression arises regularly from the meanings of its parts. For every grammatical rule determining the combination of elements, an associated rule of semantic interpretation is normally posited. According to this rule, the semantic value of a construction is computable from the values of its immediate constituents. This assumption is valid for an open-ended set of expressions. For example, the phrase *black board* is fully compositional; its content is exhausted by that of its components. The phrase refers to a board that is painted black. In spite of this, it is not difficult to find exceptions to the assumption of full compositionality, where the semantic value of an expression diverges from the predictions of regular compositional rules. This applies in particular to idioms and compounds, where the meaning of the expression tends to vary according to contextual use.

The cognitive view assumes partial compositionality. According to this assumption, the meaning of an expression is determined by both the semantic contribution of its components and the pragmatic knowledge behind what is actually symbolised. Langacker (1987: 449) writes: 'There is certainly a large measure of regularity in the meaning of composite expressions vis-à-vis the meanings of their components, and speakers rely on this regularity in determining the semantic value of novel expressions'. Yet, linguistic phenomena lend themselves more easily to a claim of partial compositionality. This is natural because language is used in context by speakers who bring many shared knowledge systems to communication. For example, the meaning of the compound *blackboard* is not fully compositional; the board is not necessarily black. It is a board of a dark colour or white for writing on with chalk or marker pen, used especially in classrooms.

In this study, the principle of compositionality is helpful to the investigation of negative words. The speaker is familiar with the intrinsic complexity of composite words. In the majority of the cases, the negative words are fully compositional, and their content resides in the components out of which they are assembled. For example, the meaning of the negative word *inaccurate* in *inaccurate information* is a combination of the meanings of the base and the prefix: not accurate, i.e. with mistakes. In a minority of cases, the negative words are partially compositional, and their content includes properties that go beyond their limits. For example, the meaning of the negative word *inhuman* in *inhuman treatment* is not only a function of the meanings of the base and the prefix not human, but rather somebody who is cruel, lacking the qualities of kindness and pity. Having said all this, one should not be misled that full compositionality is always applicable. In some areas of grammar like idioms and compounds, it fails.[6]

2.4.2 Analysability

Analysability pertains to the ability of speakers to recognise the contribution that each component structure makes to the composite whole. In Langacker's (1987: 457–460) words, it pertains to 'the extent to which speakers are cognizant (at some level of processing) of the contribution that individual component structures make to the composite whole'. As the term suggests, analysability refers to the speaker's awareness of certain aspects of a word's complexity, and thus involves cognitive events above and beyond those that constitute the word per se. This means that the speaker specifically ascribes to a composite structure like *amoral*, for instance, the contents of *a-* and *moral*, i.e. the speaker analyses *amoral* in these terms. This type of representation is referred to as *recognition*, where the components contribute to the semantic make-up of the composite structure, and so greatly enrich its interpretation. In this representation, not only the occurrence of the component parts but also their actual recognition is necessary. This is the case in which the speaker has clear intuitions about the analysability of symbolic expressions, i.e. those which are determined in context.

In this study, the principle of analysability is helpful to the investigation of negative words. The speaker recognises the contribution made by the component structures to the composite structure on the very occasion when s/he assembles them out of their smaller components. That is, the speaker is aware of the contributions made by these components. A negative word appears capable of retaining some measure of analysability; its value is enriched by the recognition of its components. When a novel expression is employed, these components are activated at some level of intensity. When a conventional expression is employed, these components need not be accessed or intensely activated on every occasion in which it is used. This is so because such expressions become entrenched or established as units. Entrenchment tends to reduce the salience of individual components; their co-activation becomes automatic. If a composite structure loses its capacity to elicit the activation of its components, it is regarded as fully opaque and unanalysable.

One consequence of analysability is that it accounts for the semantic difference observable between two expressions with essentially the same composite structure such as *dissatisfied vs. unsatisfied*, which the building-block conception fails to single out. Because component structures contribute to meaning, the non-synonymy of the expressions becomes clear. The speaker readily perceives the negative force of each prefix and recognises its contribution to the semantic value of the expression. This becomes clear when analysability is used to elicit the co-activation of the component parts of the composite structure. Another consequence of analysability is that it allows for the component structures to differ in their salience within a composite structure, which the building-block conception regards as equal. In terms of analysability, some parts are more prominent than others within a composite structure. In *amoral*, both parts figure in the expression's meaning, but the prefix *a-* is much more prominent than the base *moral*. This is so because it determines the profile of the whole expression by shifting its meaning from positive to negative.

2.5 Summary

In this chapter, I have presented a brief sketch of what morphology is and what it covers. This sketch has centred on two dimensions. One dimension is combination, the way two or more structures are joined together to form a complex structure. In this regard, I have considered two conceptions of the morphological analysis of negative words. According to the building-block conception, component structures are stacked together to form a composite structure. Conversely, a composite structure is constructed out of its components. According to the scaffolding conception, component structures serve to highlight selected facets of a composite structure. Conversely, a composite structure derives systemic motivation from its components. The other dimension is interpretation, the way in which a complex structure is understood or explained. In this respect, I have applied two principles to the morphological analysis of negative words. In the light of the compositionality principle, the meaning of a composite structure is either derived from the meanings of its components and/or includes knowledge from domains outside its limits. In the light of the analysability principle, the structure of a composite structure entails recognising the contributions made by its component structures.

To conclude the chapter, here is a summary of how derivation works from the standpoint of Cognitive Morphology:

1) In the formal view, morphology is seen as breaking up a composite word into its component subparts. The component subparts are objects being stacked together in some appropriate fashion. Their arrangement is determined by a set of rules which the language user has to follow. In *insecure*, the component substructures, the prefix *in-* and the base *secure*, function as building sub-blocks to give the overall building-block *insecure*. In the cognitive view, morphology is seen as a collection of meaningful units involving links between form and meaning. It is a network in which the language user associates the units with each other in conformity with cognitive principles. The component substructures *in-* and *secure* are semantic elements which are integrated to form the composite structure *insecure*.

2) In the formal view, a composite word is seen as a whole constructed out of smaller components, called building blocks. In the cognitive view, as Langacker (1987: 452–7) holds, a word should primarily be seen as a coherent structure in its own right, and its segmentation into components is only secondary. The components are not the building blocks out of which it is assembled, but function instead to motivate selected facets of it. Expressions correlate directly with the aspects of

meaning expressed. Bybee (1985) considers morphology as a set of lexical connections within the mental lexicon. When a word is used, its morphological elements are sequentially activated through relationships. In short, the difference is in the way the various component parts are accessed.

3) In the formal view, rules build structures and account for production. Rules produce regular patterns and decry irregular ones. In the cognitive view, schemas reflect use and account for production. The schemas which arise via repeated activation of the usage events are used to coin and understand novel expressions. The schemas subsume all regular and irregular patterns. These schemas are generalisations which are not absolute. They cover the cases that they cover, but do not deny the existence of cases which contradict them. Accordingly, there is no reason to draw a line between a module for regular or productive morphology and one for less regular or less productive morphology. Productivity, the habit of using a schema to sanction novel structures, is a matter of degree.

4) In the formal view, the construction of a word is carried out by rules, linking elements in accordance with structural parameters. In the cognitive view, the construction of a word is carried out by the language user relative to general patterns which allow for flexibility. Speakers differ in the degree to which different structures are established or conventionalised. Such variations are the crucible of language change. For example, structures differ in constituency, the order in which the components are put together to form a composite structure. Yet, the overall meaning and sound achievable through the alternate orders is the same. Likewise, structures differ in analysability, the extent to which the components contribute to the composite structure. However, the difference is just a matter of degree.

5) In the formal view, the central concern is to develop a model that describes the relationships between the subparts of a word in structural terms or categorial selection. In the cognitive view, the central concern is to develop a model that describes the relationships between the subparts of a word in semantic terms or conceptual parameters. The parameters include (i) correspondence, referring to the semantic and phonological similarity which the subparts have in common, (ii) determinacy, referring to which of the two subparts lends its profile to the entire word, (iii), dependency, referring to which of the two subparts qualifies as autonomous and which qualifies as dependent, and (iv) constituency, referring to the order in which the subparts are successively integrated to yield the entire word.

Notes

1 In the Word-and-Paradigm approach, which is alternatively called Word-based morphology, the form of a word is analysed as a whole, as the basic element of morphological description. A word is formed by defining a root within a paradigm, which is neutral with respect to the variant forms of the paradigm, and deriving the variant forms of the word from this root by using rules. The form-meaning correspondence is presumed to be many-to-one. Whereas the Morpheme-based approach treats the inflectional *-s* plural and the inflectional *-s* third person singular as belonging to one morpheme having two categories, the Lexeme-based approach assumes two separate rules for them, and the Word-based approach considers them as whole words related to each other by analogical rules. Words can be categorised based on the pattern that they fit into. The application of a new pattern can give rise to a new word such as 'older' replacing 'elder', which fits the pattern of adjectival comparatives. This approach was fundamental in Traditional Grammar. Works that fall into this Word-and-Paradigm camp include, among others, Matthews (1972) and Stump (2001).

2 Aronoff's (1976) Word Formation Rules are regular rules which operate only on words to produce new words. Every rule specifies a set of words on which it can operate. This set, or any member of this set, serves as the base of the rule. Morphemes cannot work as bases for the operation of the rules. Every rule specifies a unique phonological operation which is performed on the base. Every rule specifies a syntactic label and sub-categorisation for the resulting word, as well as a semantic reading for it. The semantic reading is the function of the reading of the base. Unlike syntactic and phonological rules, morphological rules need not be applied every time a speaker uses a word. They may be used only once.

3 Taylor (2002: 274), with whom Evans & Green (2006: 592) are in agreement, analyses English derivational prefixes, e.g. *un-*, as modifiers within a phrase, not profile determinants. This is so because they are not class-changing. They are dependent units that add additional information to an autonomous head. In direct opposition to this, I argue that they are profile determinants. It is true that they do not change the class of the bases to which they are added, but they change the meaning of the composite structure by imposing their character on the base to which they are added.

4 In the syntagmatic combination, one substructure forces a change in the specification of another substructure with which it combines. In addition to meaning, the change involves some kind of morphological and/or phonological marking. This is referred to as *coercion* and defined by Taylor (2002: 287) as: 'the phenomenon whereby a unit, when it combines with another unit, exerts an influence on its neighbour, causing it to change its specification'. In morphology, the internal structure of a word may also be obscured by coercion effects. Prefixes affect not only the semantic character of the word but determine at least some of the phonological properties of the composite word. Prefixes in English can be divided into two broad categories: those that coerce the base

to which they combine and those that do not. An example of the former is the negative-forming prefix *dis-* which changes, although very rarely, the phonological shape of the base, as in *discourage* from *encourage*. An example of the latter is the negative-forming prefix *dis-*, where the phonological shape of the base word remains unchanged, as in *dislike* from *like*.

5 In the formal view, bases and affixes differ according to structural factors. In the cognitive view, bases, or stems as Tuggy (1992: 238) calls them, and affixes differ according to other factors. (1) Bases are central to words, while affixes are peripheral, satellites to stems. (2) Affixes are bound morphemes, whereas bases are free morphemes. An affix is intrinsically incomplete and needs an independent companion. (3) Bases are semantically heavier than affixes; in that they elaborate more meanings. (4) Bases are from large, open-ended classes, whereas affixes are from closed classes. Affixes are productive, attaching to members of open-ended classes, whereas bases only attach to members of closed classes. (5) Bases have greater phonological weight than do affixes.

6 In Taylor's (2002: 116) opinion, full compositionality fails for two reasons. First, 'the conventionalised resources of a language are abstractions over usage events. Their semantic content encapsulates common features of their many uses, filtering out the specifics of individual events. Even when a ready-made expression is applied to a situation, the specifics need to be filled in again'. Second, when component substructures come together, they need to adjust to each other in certain details. This requires shifting their values relative to the intended conceptualisation. This process is referred to by Langacker (1991: 543) as *accommodation*: 'the adjustment a component structure undergoes when integrated with another to form a composite structure'. For example, the meaning of the word *human* has both positive and negative qualities. In *inhuman*, the root *human* accommodates itself to the meaning of the prefix *in-* to denote cruelty. In Cognitive Semantics, it is claimed that linguistic units do not normally have fixed meanings which they contribute to the composite expression in which they occur. Just as the articulation of a phonetic segment adjusts to the articulation of an adjacent segment, so a contributing semantic unit typically varies according to the units with which it combines.

3 Category

3.0 Overview

This chapter explores the first axis in my semantic system, i.e. the role of categorisation in the semantic description of negative prefixes as single lexemes. To do this, I take my cue from two tenets of Cognitive Semantics. One tenet is that all linguistic items, lexical or grammatical, have semantic values which motivate their linguistic behaviour. Applying this tenet to morphology, I argue that a prefix has semantic import and adds to the meaning of the base which hosts it. Another tenet is that linguistic items are polysemous by nature and so form complex categories. Applying this tenet to morphology, I argue that a prefix forms a category made up of a wide range of senses, which are organised around a central one, termed the *prototype*. The other senses, termed peripheral, are derived by the interaction between the semantics of the prefix and the base. To this end, the chapter is organised as follows. Section 1 delineates the phenomenon of polysemy and introduces the questions raised in its investigation. Section 2 discusses two theories of linguistic meaning. One is the classical theory which comprises homonymy and monosemy. The other is the modern theory which comprises family resemblance and prototypes. Section 3 applies the assumptions of the *prototype* theory to the description of negative prefixes. Section 4 provides a schema for each of the negative prefixes. Section 5 gives a summary of the chapter and reaches a number of conclusions.

3.1 Introduction

One phenomenon that is endemic in the lexicon of any natural language pertains to *polysemy*, that is, where a linguistic item, usually a *lexeme*, has more than one meaning. The lexeme *head*, for instance, has at least 5 senses in the *Online Cambridge Advanced Learner's Dictionary*. The lexeme is the fundamental unit of vocabulary. It is well known that language elements, lexical items or grammatical morphemes, have a wide range of meanings, which are manifested in the contexts in which they are used. In the literature, there has been a great deal of discussion of the semantic analysis of such elements. While there may be agreement on the phenomenon itself, its analysis has proved to be very controversial. All theories agree on the multiplicity of senses of a given lexeme, but they differ in how to represent them. The controversy, so to speak, centres on the question of how one treats the many senses attributed to a given lexeme. Linguists with different theoretical affiliations give different accounts of how to describe the semantic profile of a given lexeme.

In grappling with the phenomenon of polysemy, linguists come up against two central questions. The first question is: how are the senses of a lexeme related to one another? To put it differently, is there a connection of some sort between the senses of a lexeme? The second question is: what principles govern the relationship between the senses? In other lexemes, is the relationship between the senses motivated? The present study attempts to find answers to such questions, and so provide an account of how lexemes in natural language behave and how they are organised. To do so, this chapter is structured as follows. Section 2 looks at some of the theory-related stances concerning the lexical profile of a lexeme. It is divided into two parts. Part one addresses the classical theory, and part two addresses the modern theory. Section 3 forms the heart of the chapter. It extends the prototype theory to the analysis of negative prefixes, which the study claims holds sway over the other theories. Section 4 presents detailed descriptions of the semantic networks of the negative prefixes.

3.2 Theories of categorisation

Speakers describe their experiences of the world not only in terms of individual things but also in terms of categories of things. Any category is a set of conceptual experiences or a class of linguistic expressions which have properties in common. Categorisation is central to human thinking. Whenever speakers reason about an entity, they categorise it as a kind of thing. Entities may be abstract, such as events, actions, emotions, etc. or may be concrete such as chairs, houses, trees, etc. Without the ability to categorise, humans could not function at all. Categorisation is the process of grouping together conceptual experiences or linguistic expressions into a category. Categorisation is the most fundamental of human capacities, which is essentially a matter of both perception and imagination. In the realm of language, the idea of a category is prevalent, too. Humans have the capacity for categorising lexical items. This is because most lexical items do not designate individual things, but rather sets of things that are joined together by commonality.

To address the issue of the meaning of a given lexeme, linguists have developed two theories: classical and modern. Both theories consider categories a pivotal means of making sense of experience. Yet, they stand in stark contrast to each other regarding how people categorise and on what grounds they do so. The former has an affinity to philosophy and logic. It relates meaning to truth conditions and describes the meanings associated with an entity in terms of features. The latter has an affinity to psychology. It relates meaning to mental representations or bodily experiences, and describes the meanings associated with an entity in terms of imagery. In connection with language, the classical theory places the regular patterns in the syntax, and delegates the irregular ones to the lexicon, whereas the modern theory places all regular and irregular patterns in the lexicon. In what follows, I provide an overview of the stances held by the different theories of meaning. For an extensive overview of the different theories, see Lakoff (1987: xi-xvii), Geeraerts (1988: 652–7), Taylor (1989: 21–37), Ravin & Leacock (2002: 1–29) and Evans (2003: 79–104).

3.2.1 The classical theory

The classical theory on categorisation can be traced back to Aristotle. The main thrust of this theory is that it defines a category in terms of a set of defining features. The features are singly necessary and jointly sufficient for an entity to be a member of a category, and each feature is equally necessary for membership in the category. If an entity has all of the necessary features, the entity is included in the category. If the entity does not have one or more of these features, the entity is excluded from the category. The boundaries are clearly defined, so membership in a category is clear-cut. Accordingly, each member is equally representative of the category. For example, an entity belongs to the category X if and only if it possesses all the features which define the category. The lack of any of the defining features removes the entity from the category. Membership is an either-or matter. Features are either present or absent; they cannot apply only to some degree. An entity either is or is not a member of a category. It cannot be a quasi-member of a category.

The classical theory hinges, as summarised by Taylor (1989: 23–24), on four assumptions. First, categories are defined in terms of a conjunction of necessary and sufficient features. Second, members of a category share the same set of definitional features. Features are a matter of all or nothing. A member either possesses this feature or it does not. A feature is either involved in the definition of a category, or it is not. Third, categories have clear boundaries. A category, once established, divides the universe into two sets of entities: those that are members of the category, and those that are not. There are no middle cases, where some entities in some way belong to the category but in another way do not. Fourth, members of a category have equal status. Any entity which exhibits all the defining features of a category is a full member of that category. An entity which does not exhibit all the defining features is not a member. There are no degrees of membership in a category, i.e. there are no entities which are better members of the category than others.

In the area of language, the classical theory views the mental lexicon as a repository of irregularity, and syntax as a container of regularity. In constructing a category for a lexical item, the senses that are regular are included, whereas the senses that are irregular are excluded and transferred to the lexicon. This is so because categorisation undergoes processes that are productive, rule-governed and predictable. Chomsky (1995: 235) writes: 'I understand the lexicon in a rather traditional sense: as a list of 'exceptions', whatever does not follow from general principles'. Models adopting this theory have tended to represent different lexeme senses as distinct lexical items. Polysemy is simply represented as consisting of an arbitrary list of discrete lexemes that happen to share the same phonological form. As Jackendoff (1997: 4) contends, the lexicon has been represented as a static set of lexeme senses, tagged with features for syntactic, morphological and semantic information, ready to be inserted into syntactic frames with appropriately matching features. The lexicon has thus been viewed as a finite set of discrete memorised units of meaning.

An embodiment of the classical theory is the rationalist theory of semantics developed by Katz & Fodor (1963) and later refined in Katz (1972). According to this theory, the different senses of lexical items are represented by semantic units, called *Semantic Markers*, which correspond roughly to Aristotle's defining features. A difference in a semantic marker is enough to distinguish one sense from another. Senses can be represented independently of the context in which they occur. For example, one sense of *chair* would be represented by markers such as: *object, physical, non-living, artefact, furniture, portable, having legs, having a back, having a seat*, etc. However, as Ravin & Leacock (2002: 9) contend, Katz's theory does not distinguish polysemy from homonymy, as it is not the goal of his theory to explain the etymology of sense differences. Similarly, Katz's theory does not explain regular polysemy, where a lexical item acquires different senses extended in predictable ways.

The classical theory of linguistic categorisation can be narrowed down to two approaches. One approach locates lexical meaning in homonymy, whereas the other approach locates it in monosemy.

3.2.1.1 Homonymy

According to the homonymy position, the different meanings associated with a lexical item are considered odd, with each meaning being completely unrelated to the other. The basic idea is that it is merely an accident that different meanings happen to be attached to a lexical item. For example, the meanings associated with the lexical item *head* are unrelated to one another although they share the same form. This position, as assessed by Evans (2003: 87–8), has two significant problems. First, it denies systematicity in the relationship between the different meanings of an item. It ignores the fact that lexical organisation is highly structured, as evidenced by Lakoff (1987), Brugman & Lakoff (1988), Langacker (1987, 1991), Levin (1993) and Tyler & Evans (2003). Second, it denies the principled nature of language change. It ignores the fact that lexical meaning constitutes an evolving system whose changes are motivated, as evidenced by Sweetser (1990), Hopper & Traugott (1993), Bybee et al. (1994), Heine (1997) and Riemer (2005). For present purposes, the homonymy position hence fails to account for the different meanings associated with a lexical item.

3.2.1.2 Monosemy

According to the monosemy position, a lexical form pairs with a single sense. The various meanings associated with a lexical item can simply be explained in terms of contextually derived variants of a single sense. In Ruhl's (1989) theory, the different meanings associated with a form are derived from it by contextual knowledge. In Pustejovsky's (1995) theory, the different meanings are derived by means of a series of generative devices from an underspecified abstract representation. Whereas Ruhl's theory values context in producing meanings, Pustejovsky's theory values parsimony in producing meanings. This position, as appraised by Evans (2003: 89), has its own limitations. First, there are some meanings which are context-independent. Although contextual knowledge is important in associating different meanings with a lexeme, it is in some cases insufficient in predicting the conventional meanings ordinarily derived. Second, it is true that different meanings can be derived from an abstract representation, but they are not argued or justified, as evidenced by a variety of papers edited in Nerlich (2003) and Cuyckens (2003). For present purposes, the monosemy position thus fails to account for the semantic behaviour of a lexical item.

In conclusion, the classical theory is not helpful for the semantic analysis of lexical items. While the phenomenon of polysemy is not a major problem for the classical theory, sense distinctions apparently are. The semantic content of a lexeme can be captured by a finite set of necessary and sufficient features. For the meaning of a lexeme, one has to find the common element in all its applications. Lexical meaning is reducible to a finite set of necessary and sufficient features. Such assumptions can be called into question. First, it considers all features for inclusion in a category as equally salient. It ignores the fact that some features may appear more salient than others. Second, it rates all members of a category as equally representative. It ignores the fact that some members of a category may appear more typical than others. Third, it regards the boundaries as clear-cut. For all practical purposes, it is difficult to determine to which category a given entity belongs. Further, it is difficult to specify what the defining features are for a given category. In a well-known discussion of the meaning of the lexeme 'game', Wittgenstein examined *board games, card games, ball games* and *Olympic games* and failed to find an element that is common to all.

3.2.2 The modern theory

In the classical theory, on whose foundations Formal Semantics is built, the description of lexical meaning is carried out in terms of defining features. In response to it, the modern theory describes lexical meaning in terms of either a chain of resemblances or a prominent example. Accordingly, it subsumes two approaches. One involves the concept of *family resemblance*. The other involves the concept of *prototype.*

3.2.2.1 Family resemblance

In view of Wittgenstein's (1953, 1958) *family resemblance* approach, the different entities which form a category are treated like members of a large family. They don't share defining features, but rather family resemblances which overlap in the same way attributes do in families. The entities resemble one another in a variety of ways. No attribute is shared by all the entities of a category. There may be entities which share no attributes, yet they are linked to the others by a chain of resemblances. No entity combines all attributes defining membership within a category. That is, it is not necessary for an entity to have all the attributes which allow it to be a member of a category. Each entity has at least one and probably more attributes in common with the other entities. No or few attributes are common to all the entities in the category. Although it was important in breaking the stranglehold of the classical theory, this theory encounters a criticism too. Wittgenstein does not say what the attributes are, and how in-family attributes differ from out-family ones.

3.2.2.2 Prototype

In the light of Rosch's (1977, 1978) *prototype* approach, a category is defined by an intersection of properties. The member that has all of the properties is the prototype, i.e. the ideal or central member. Members that contain some, but not all, of the properties are peripheral. The different members of a category are not equal. They are assimilated to the category or not, according to what degree they resemble the prototype. They conflict in some ways with the specifications of the prototype, but are assimilated to the category on the basis of some perceived similarity to the prototype. That is, the members are categorised not on the basis of necessary and sufficient conditions, but rather on the basis of resemblance to the exemplary member. Since the boundaries between the members of a category are fuzzy, the structure of a category takes the form of overlapping properties. Category membership is a gradient rather than an absolute. It is a matter of degree in the sense that it is based on how similar the properties of a member are to those of the prototype.

Geeraerts (1988: 654–5) gives a summary of the main assumptions of the *prototype* theory. First, categories are not defined by a set of necessary properties which can be predicated of the members of the category in question. Some of the allegedly necessary properties of the central example may appear to be optional at the periphery. Second, properties within a category have different, rather than equal, degrees of salience. Properties not shared by all the members are less important than properties that appear in all or most of the members. Third, category boundaries are vague rather than clear. A category contains marginal instantiations that do not conform rigidly to the conceptual centre. Fourth, category membership is defined by similarity rather than identity. Category membership is not a question of either-or, but a matter of degree. Therefore, not every member is equally representative of a category.

In the area of language, the modern theory takes a significantly different view of the nature of the mental lexicon. The mental lexicon does not represent lexemes as bundles of semantic, syntactic and morphological features, but rather as imagistic, schematic representations. That is, the meanings associated with a lexeme are instantiated in memory not in terms of features, nor as abstract propositions, but rather as conceptual networks. As Johnson (1987) asserts, such meanings are held to be embodied in the sense that they arise from perceptual reanalysis of recurring patterns in everyday physical experience. A number of linguists such as Rosch (1977), Fillmore (1982), Brugman (1988), Levin (1993) and Lakoff (1987) applied the theory to the human lexicon and found that lexical items constitute natural categories involving a network of related senses, which are organised with respect to a primary sense. Hence, such accounts are strongly suggestive that the lexicon is much more motivated and organised than has classically been assumed.

3.3 Negative prefixes

In the present analysis, I extend the *prototype* theory to morphology. I seek to show how relevant the concept of prototypes is to the analysis of a negative prefix, and how useful such an analysis is in resolving questions concerning the emergence of multiple senses, their extensions and the relatedness among them. On a general basis, I follow Langacker (1987: 416, 1991: 35) who writes, 'a frequently-used morpheme or lexical item has a variety of interrelated senses'. Likewise, I follow Fillmore & Atkins (1992) who view lexical meanings as networks of semantic concepts that are extensible from a core. The direction and scope of the extensions depend on the nature of the lexical item analysed and the context in which it is used. Applying this to the area of prefixation, I argue that a negative prefix forms a category having a prototype, with the other senses gathering around it on the basis of semantic similarity. The speaker's choice of sense is based on the way s/he conceptualises a situation and the morpheme that s/he selects to represent it.

On a specific basis, I build on my own work (Hamawand, 2007). At the synchronic level, a negative prefix is made up of distinct senses which are structured in the shape of a network. The senses are related to one another in a systematic way. They are the result of the dynamic process of meaning-extension, which is a function of language use and the nature of socio-physical experience. The senses are organised with respect to a central, prototypical, sense. The central sense is the sense that comes to mind first when thinking of the meaning of the prefix, occurs most frequently and is the most basic in its capacity to clarify the other senses. The other senses range over a continuum from less prototypical to peripheral. They are modelled in terms of relative distance to the central sense. Accordingly, the more peripheral senses are less related to the central sense than the less peripheral ones. The borders between the senses within the network are extremely fuzzy or unclear, and so the senses tend to overlap with one another in meaning.

3.3.1 Assumptions

To account for the internal structure of a negative prefix, the present study draws on insights from Cognitive Linguistics. Based on such insights, the study makes two assumptions:

1) The lexicon is not an arbitrary repository of unrelated lexemes. Rather, the lexicon, as Langacker (1988a) claims, constitutes a network of form-meaning associations. Grammar is inherently iconic in that there is a non-arbitrary relationship between structure and content. As Givón (1990: 966–76) explains, grammatical structure is somehow isomorphic to its function. Autonomous approaches maintain a strict separation between the realms of form and meaning. As a result, they basically ignore the influence of one upon the other. To them, structure is independent of content. In this study, my intention is to show that an iconic relationship between form and meaning can be demonstrated for the multiple uses of a negative prefix across an array of examples in

which it occurs, with the result that the different examples reflect subtle differences in meaning. This follows the cognitive tenet that linguistic structure reflects semantic structure, as manifested in Fauconnier (1985), Heine (1997), Lakoff (1987) and Langacker (1987,1991).

2) The lexemes form networks in the mental lexicon. A single lexeme is associated with a large array of senses related via a variety of semantic principles and arranged with respect to a primary sense. The different senses are best represented as a complex category, in which one sense serves as a standard from which the other senses are derived via semantic extensions. As Langacker (1987: 110–111) claims, different morphosyntactic constructions are assumed to reflect different conceptualisations, and thus to have different meanings. In this study, my intention is to show that the meaning of a negative prefix can be characterised by a complex network of interrelated senses. The senses of a negative prefix are semantically motivated and related to each other. This follows the cognitive tenet that human conceptualisations derive from human perceptions and experience of the spatio-physical world, as demonstrated by Clark (1973), Lakoff & Johnson (1980), Sweetser (1990) and Talmy (1983).

3.3.2 Goals

To date, neither has the internal structure of a negative prefix been wholly addressed nor have its semantic contributions been comprehensively explained. As Atkins & Levin (1991) stress, there is little agreement among lexicographers as to the number of senses they register for a lexeme, the ways in which the different senses are organised, the principles that govern the senses, and the definitions they provide. To account for the meanings of the negative prefixes, the present study adopts the assumptions used within the *prototype* theory. Based on such assumptions, the study aims to:

1) Demonstrate that language resides in the minds of its speakers, not in dictionaries. In order to understand the nature of language, we have to look at our conceptual world and how it shapes the signs. Language helps speakers to categorise their experiences of the world. The conceptual system is organised in terms of categories. The mind is imaginative; it is more than a mere mirror of nature or a processor of symbols. The goal is to show how the meanings of a negative prefix relate to one another and to the entities existing in the conceptual world.

2) Explicate that meanings are not represented in the mind as single entities, but rather as a network of distinct but related senses. The different senses associated with a lexical item are structured round a centre. The senses are related to one another in principled ways. The senses are not related to one another in equal ways. Some may be directly related to the centre; others may be indirectly related to the centre. The goal is to show that the semantic network of a negative prefix has a clear centre, but its fuzzy boundaries embrace overlap.

3) Indicate that lexical items function in a flexible manner rather than in a rigid fashion. The representation of the senses of a lexical item is a highly motivated process, which derives new senses from the existing ones. Different types of factors operate on the senses to yield different interpretations in different contexts. It is quite normal in language that a lexeme acquires different readings in different contexts. The goal is to pinpoint the semantic principles which determine the interaction between meaning and context in a negative prefix.

3.3.3 Procedures

Polysemy, as Langacker (1988a: 50–2) emphasises, is the norm for lexical items, and so is a natural phenomenon. Any negative prefix forms a network of multiple senses. A speaker's knowledge of a given prefix usually embraces its entire network. The form of the network may vary from speaker to speaker depending on their experience and their categorisation. The senses differ in terms of salience, how entrenched they are; value, how conflicting they are; and distance, how far they are from the basic sense. To fulfil the goals outlined above, the study follows such routes as:

1) Determining the central sense of a negative prefix. There are some crucial factors which help to determine the most central sense of a given prefix. The central sense is the one that comes to mind first. It is the sense that occurs most frequently in our experience. It is the sense that stands out as the most salient among the category senses. It is the sense that is most basic in its capacity to clarify the other senses. It is the sense that serves as a reference-point for the less salient senses.

2) Identifying the marginal senses of a negative prefix. There are some key principles which provide a means of identifying the distinct senses instantiated in the semantic network for a given prefix. A distinct sense is the one that has an additional meaning that is not apparent in the other senses. It is the sense that is derived from context or inferred from

another sense. It is the sense that displays certain structural properties. It is the sense which is rated according to how similar it is to the prototype.

3) Laying down the itinerary which links the senses. There are some pivotal means which help to clarify the relations between the different senses of a negative prefix. The linkage is motivated by semantic rather than syntactic considerations. Such considerations make the transition possible from one sense to another. Any negative prefix has a core sense; the other senses arise because they satisfy different semantic conditions. This process helps to uphold the unity of the negative prefix.

3.4 Prefixal networks

Primarily, a negative prefix indicates a state or action that is opposed to the base to which it is attached. Secondarily, it is a morpheme which is added to a semantically positive or neutral base to signal adverseness. The attachment of a prefix to a base is affected by some parameters. The first pertains to correspondence. Phonologically, the negative prefix *in-*, for example, makes a schematic reference to a base, e.g. *correct* elaborates to yield *incorrect*. Semantically, the negative prefix *in-* denotes the relation of distinction which corresponds to the relation denoted by *correct*. The second relates to dependence. The negative prefix *in-*, for example, qualifies as dependent (a bound morpheme), and so cannot exist on its own. To complete its meaning, it needs a base like *correct*, which qualifies as autonomous (a free morpheme) to form *incorrect*. The third belongs to determinacy. The negative prefix *in-*, for example, means 'not' and determines the character of the entire word *incorrect* to mean 'not correct'. The fourth concerns constituency. In *incorrect answer*, at the first level of constituency, the base *correct* integrates with *in-* to form *incorrect*. At the second level, *incorrect* combines with *answer* deriving thus the composite structure of the expression.

Before turning to the characterisation of the negative prefixes, a word on the morphological structure of a negated word is appropriate. The present study covers only negative words meeting three conditions. First, a negated word should have a corresponding positive counterpart. For example, a word like *uncommon* is included as it has the positive form *common*. By contrast, a word like *unbeknown* is not included as it does not have a positive form. Second, the meaning of the negative and positive forms should be semantically related. For example, words like *uncommon* and *common* are semantically related because *uncommon* means 'not common'. So, such words are included. By contrast, words like *indifferent* and *different* are not semantically related since *indifferent* does not mean 'not different'. Such words are therefore excluded. Third, the lexical item to which the prefix is attached must be a free base. For example, a word like *uncommon* can be separated into *un-* and the free base *common*. Accordingly, such words are included. By contrast, a word like *dissociate* is opaque because it does not have a free base, as *sociate* does not exist on its own in English. Accordingly, such words are excluded.

The meaning of a negative prefix is established in terms of semantic principles. When the base is a noun, the prefix is defined relative to either animacy or concreteness. Animacy has two elements. One describes an animate entity, whereas the other describes an inanimate entity. Concreteness also has two ingredients. One describes a concrete entity, while the other describes an abstract entity. When the base is an adjective, the prefix is defined with reference to property, which also has two constituents. One describes quality, whereas the other describes quantity. In addition, the prefix is defined with reference to gradability. When the base is gradable, the prefix forms a negative that is contrary in meaning. *Contrary* here means that the base denotes a quality which can be present in varying degrees: that is, linguistically it can be modified by adverbs like 'very'. When the base is non-gradable, the prefix forms a negative that is contradictory in meaning. *Contradictory* here means that the base denotes a quality which admits of no middle ground: that is, linguistically it cannot be modified by adverbs like 'very'. Finally, when the base is a verb, the prefix is defined with reference to transitivity. Transitivity has two components. One is transitive, where the action requires a patient. The other is intransitive, where the action does not require a patient.

3.4.1 Primary negative prefixes

Primary negative prefixes are bound morphemes used to indicate the opposite of the base to which they are attached. They include *a(n), de-, dis-, in-, non-* and *un-*. Below is a prototype-based characterisation of each.

3.4.1.1 a(n)-

The negative prefix *a(n)-* is borrowed from Greek. The *n* appears before bases beginning with a vowel, as in *anaesthesia*. The prefix is found almost exclusively with words formed from Greek bases. In forming negative words, the prefix *a(n)-* is class-preserving.[1]

Prototypically, the prefix *a(n)-* is attached to adjectival bases to express distinction. Depending on the nature of the combining base, its meaning can be paraphrased as:

a. 'divergent from the quality referred to by the adjectival base'. This meaning arises when the prefix is attached to gradable adjectival bases, which describe humans. For example, *an amoral instigator* is an instigator who does not have moral principles, *an apolitical student* is a student who does not have interest in political matters, and *an asocial man* is a man who does not have social skills or connections.

b. 'unlike the quality referred to by the adjectival base'. This meaning arises when the prefix is attached to non-gradable adjectival bases, which describe non-humans. For example, *an ahistorical phenomenon* is a phenomenon that is not related to history or tradition, *an asymmetrical shape* is a shape that is not symmetrical or balanced, and *an atypical pattern* is a pattern that is not typical or usual. Further examples include *aseptic, atemporal, atonal*, etc.

Peripherally, the prefix *a(n)-* is attached to other bases to express other meanings. Depending on the nature of the combining base, its meaning can be paraphrased as:

a. 'without the thing referred to by the adjectival base'. The meaning of privation arises when the prefix is attached to non-gradable adjectival bases derived from nouns, which describe entities. It is chiefly used in medical terms. For example, *acardiac* means without a heart, *acaudal* means without a tail, *aglossal* means without a tongue, and *asexual* means without sexual organs, without sex or without sexual desire. In a few cases, the prefix applies to descriptions of inanimate entities. For example, *achromatic* means without colour, and *anhydrous* means without water.

b. 'not adhering to the belief referred to by the nominal base'. The meaning of opposition arises when the prefix is attached to nominal bases, which imply abstraction. For example, *an atheist* is a person who does not adhere to theism, the belief that there is a God; *an agnostic* is a person who does not adhere to gnosticism, the belief that God's existence is provable; and *an anarchist* is a person who does not adhere to archism, the belief that calls for a system of government in society. In modern English, the positive forms of these words are either rarer than their negative counterparts or obsolete.

Figure 7 presents a graphical representation which captures the multiple uses of the negative prefix *a(n)-*. Note that the solid arrow represents the prototypical sense, whereas the dashed arrows represent the semantic extensions.

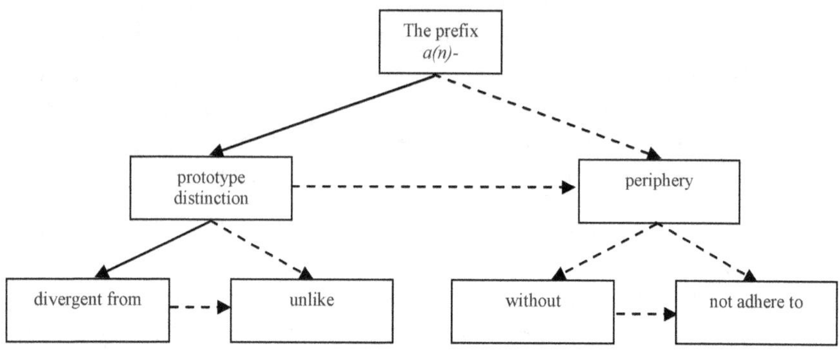

Figure 7: The semantic network of the negative prefix *a(n)-*

3.4.1.2 *de-*

Prototypically, the prefix *de-*, which is derived from Latin, is annexed to nominal bases to form verbs. Relying on the nature of the integrated base, the prefix *de-* displays the following semantic patterns:

a. 'reversing the action described by the nominal base'. The meaning of reversal is construed when the negative prefix is affixed to nouns denoting things or places. In some cases, the prefix is annexed to simple nouns. In this use, the prefix is class-changing. For example, *decontrol* means freeing restraints from something, *deface* means spoiling the appearance of something, and *deform* means changing the natural shape of something. A list of other examples includes *decipher, decode, defile, deforest, derail*, etc.[2]

   In other cases, the prefix is affixed to verbs derived from nouns by means of the suffixes *-ate, -ify* and *-ise*. In this use, the prefix is class-preserving. For example, *declassify* means stating officially that secret government information is no longer secret, *deregulate* means freeing a trade or a business activity from rules and controls, and *destabilise* means making a system, country or government become less firmly established or successful. A list of other examples includes *deactivate, decentralise, decontaminate, de-escalate, dematerialise, demilitarise, depolarise, desegregate*, etc.[3]

b. 'removing the thing described by the nominal base'. The meaning of removal is construed when the negative prefix is affixed to concrete nouns denoting non-humans. In this use, the prefix is class-changing. In some cases, the prefix triggers the process of removal, and so serves as a causative element. For example, *debug* means removing defects in a device, system or plan, *defuse* means removing the fuse from a bomb so that it cannot explode, and *demist* means removing the condensation from a car's windows. A list of other examples includes *defog, defrost, degas, degrease, de-ice*, etc.

   In other cases, the prefix reiterates the idea of removal, and so serves as an emphatic element or an intensifier. For example, *debone* means removing the bones from meat or fish, *delouse* means removing lice from something, and *deworm* means removing worms from something. A list of other examples includes *deflesh, dehair, dehull, denude, descale*, etc.

c. 'depriving of the thing described by the nominal base'. The meaning of privation is construed when the prefix is affixed to simple noun bases denoting things or places. More precisely, it means stripping or ridding something of a certain quality. In this use, it is class-changing. For example, *deface* means depriving something of its true shape by writing on it, *dehydrate* means depriving something, especially food, of water, and *despoil* means depriving a place of something valuable. A list of other examples includes *debase, decolour, deform, etc.*

Peripherally, the prefix *de-* is annexed to nominal bases to express other meanings. Relying on the nature of the integrated base, *de-* displays the following semantic patterns:

a. 'reducing the thing described by the nominal base'. This meaning is construed when the prefix is affixed to abstract nouns denoting non-humans. For example, *debase* means lowering the quality, character or value of something, *degrade* means reducing the grade of something, and *devalue* means reducing the value of something. A list of other examples includes *declass, decompress, depress,* etc.

b. 'analysing the thing described in the nominal base'. This meaning is construed when the prefix is affixed to abstract nouns denoting non-humans. For example, *decode* means finding the meaning of something, especially something that has been written in code, *deconstruct* means analysing a text in order to show that there is no fixed meaning within the text but that the meaning is created each time in the act of reading, and *demystify* means making something easier to understand and less complicated by explaining it in a clear and simple way.

c. 'getting off the vehicle described by the nominal base'. This meaning is construed when the prefix is affixed to concrete nouns denoting humans. Precisely, it means disembarkation. For example, *debus* means alighting from a bus, *deplane* means getting off a plane, *detrain* means leaving a train or make somebody leave a train.

d 'cancelling the thing described in the nominal base'. This meaning is construed when the prefix is affixed to nouns. For example, *decommission* means stopping the use of weapons, and *decouple* means cancelling the relationship between two things. In rare cases, the base is verbal as in *deselect* which means cancelling the selection of an existing member of parliament for re-election.

Figure 8 presents a graphical representation which captures the multiple uses of the negative prefix *de-*. Note that the solid arrow represents the prototypical sense, whereas the dashed arrows represent the semantic extensions.

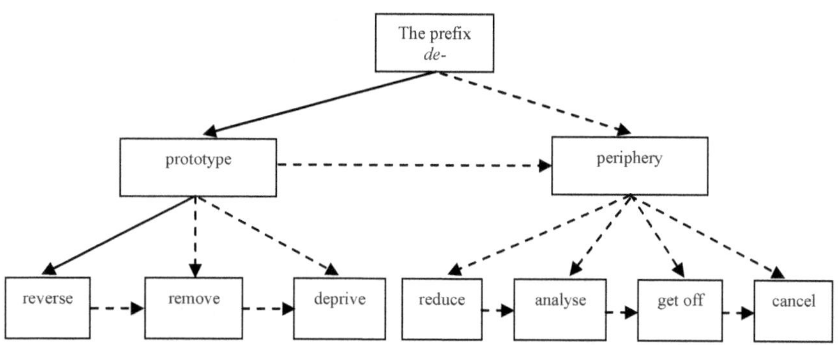

Figure 8: The semantic network of the negative prefix *de-*

3.4.1.3 *dis-*

The negative prefix *dis-* is from Latin. It is sometimes spelled *dys-* in words typically confined to medical contexts, as in *dysentery*. Like the other negative prefixes, *dis-* allows a broad selection of bases.

Prototypically, the prefix *dis-* is combined with adjectival bases to express distinction. It is class-preserving. Relative to the nature of the combining base, the prefix expresses the following senses:

a. 'the converse of the quality signified by the adjectival base'. This subtlety of meaning follows when the prefix is combined with adjectival bases implying a quality. The new formations are gradable and imply contrary opposition. For example, *disloyal* is the converse of loyal, *disobedient* is the converse of obedient, and *dispassionate* is the converse of passionate. A collection of other examples includes *disaffected, discourteous, dishonest, disingenuous, disobliging, disproportional, disreputable, dissimilar,* etc.

b. 'unwilling to consider the state signified by the verbal base'. This subtlety of meaning follows when the prefix is combined with verbal bases implying a state. For example, *disbelieve* means to be unwilling to believe somebody or something, *dislike* means to be unwilling to like somebody or something, and *distrust* means to be unwilling to trust somebody or something. A collection of other examples includes *disaffect, disapprove, disown, displease, disremember, etc.*

Peripherally, the prefix *dis-* is combined with other bases to express other meanings. Relative to the nature of the combining base, *dis-* expresses the following senses:

a. 'turning around the action signified by the verbal base'. The meaning of reversal follows when the prefix is combined with verbal bases implying change.[4] In this use, the prefix is class-preserving and affects people. For example, *discharge* means to allow somebody to leave prison or a court of law, *disengage* means to free somebody from the thing that is holding them, and *disqualify* means to prevent somebody from doing something because they have broken a rule or are not suitable. A collection of other examples includes *disband, discredit, disempower, disentangle, disinherit, displace, displease, dissociate*, etc.

b. 'ridding of the thing signified by the nominal base'. The meaning of removal follows when the prefix is combined with concrete nouns. In this use, the prefix is class-changing and affects people. For example, *disarm* means to rid somebody of weapons, *disenfranchise* means to take away somebody's rights, especially their right to vote, and *disrobe* means to rid somebody of clothes, especially worn for an official ceremony. In some formations, the bases are verbal. For example, *dishearten* means making somebody lose hope or confidence, *dishonour* means making somebody or something lose the respect of other people, and *dispossess* means making somebody lose property, land or house.

c. 'lacking the thing signified by the nominal base'. The meaning of privation follows when the prefix is combined with abstract nouns. In this use, the prefix is class-preserving and affects people. For example, *disbelief* means a lack of belief, *discourtesy* means a lack of courtesy, and *disrespect* means a lack of respect. A collection of other examples includes *disgrace, discomfort, disfavour, disharmony, dishonour, disobedience, disrepute, disservice*, etc.

Figure 9 presents a graphical representation which captures the multiple uses of the negative prefix *dis-*. Note that the solid arrow represents the prototypical sense, whereas the dashed arrows represent the semantic extensions.

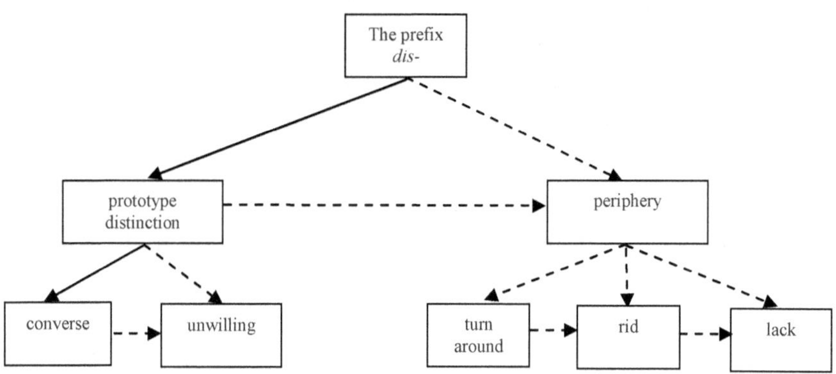

Figure 9: The semantic network of the negative prefix *dis-*

3.4.1.4 *in-*

The negative prefix *in-* is of Latin origin, so it is usually seen when the base is from Latin. Before certain letters, the *n* changes to another letter: *il-* before *l* as in *illogical*, *im-* before *b*, *m*, or *p* as in *impossible*, and *ir-* before *r* as in *irregular*. In the new formations, the prefix *in-* is class-preserving.[5]

Prototypically, the prefix *in-* is linked to adjectival bases to express distinction. Relying on the nature of the combined base, *in-* displays the following semantic functions:

a. 'the opposite of the property expressed by the adjectival base'. This meaning succeeds when the prefix is linked to simple adjectives, which are gradable. In this function, the prefix denotes properties of situations. For example, *inaccurate* is the opposite of accurate, *indiscreet* is the opposite of discreet, and *impossible* is the opposite of possible. Examples of other words are *inactive, inanimate, inappropriate, incomplete, illegal, improbable, irrational, insecure, insignificant, irregular*, etc.

b. 'difficult to perform the process expressed by the adjectival base'. This meaning succeeds when the prefix is linked to adjectives ending in *-able* or *-ible*, which are gradable. In this function, the prefix negates a process. For example, *incomprehensible* means difficult to understand, *incurable* means extremely difficult to cure, and *inedible* means difficult to eat or cannot be eaten. Examples of other words are *inadmissible, incalculable, inexpressible, implausible, incredible, inviolable, insurmountable, intangible, intolerable, invisible,* etc.[6]

Peripherally, the prefix *in-* is linked to nominal bases to express other meanings. Relying on the nature of the combined base, *in-* displays the following semantic functions:

a. 'empty of the thing expressed by the nominal base'. The meaning of privation succeeds when the prefix is linked to abstract nouns denoting state. For example, *ineptitude* means a lack of skill, *imbalance* means a lack of sameness, and *insanity* means a lack of reason. Examples of other words are *inaptitude, incompetence, inequality, infelicity, ingratitude, inhumanity, injustice, insignificance*, etc.

b. 'being unable to do the action expressed by the nominal base'. The meaning of inability succeeds when the prefix is linked to abstract nouns denoting action. For example, *inaction* means an inability to act, *incomprehension* means an inability to understand somebody or something, and *indecision* means an inability to decide. Examples of other words are *inattention, indiscipline, indisgestion, indisposition, inexperience*, etc.

Figure 10 presents a graphical representation which captures the multiple uses of the negative prefix *in-*. Note that the solid arrow represents the prototypical sense, whereas the dashed arrows represent the semantic extensions.

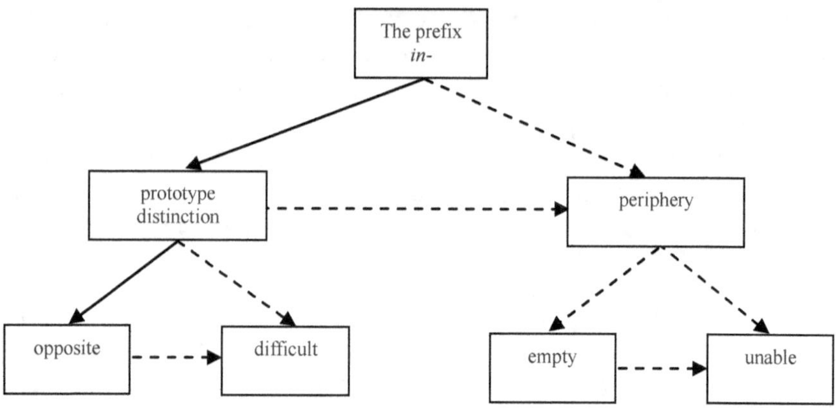

Figure 10: The semantic network of the negative prefix *in-*

3.4.1.5 *non-*

The negative prefix *non-* is from Latin. As *non-* is a productive prefix, the list of words, technical and non-technical, hosting it is practically unlimited. In all formations, *non-* is class-preserving.[7]

Prototypically, the prefix *non-* is connected with nominal bases to express distinction. It shows such subtle differences in meaning as:

a. 'failing to do the action described by the nominal base'. This nuance appears when the prefix is connected with suffixed abstract nouns denoting human actions. Precisely, it indicates refusal or failure. For example, *non-compliance* refers to the failure or refusal to obey a rule, *non-payment* refers to the failure to pay a debt, tax or rent, and *non-proliferation* refers to the limitation of the production or spread of something, especially nuclear or chemical weapons. Other words include *non-adherence, non-acceptance, non-appearance, non-cooperation, non-interference, non-intervention, non-obedience, non-resistance*, etc.

b. 'not fulfilling the requirement described by the nominal base'. This nuance appears when the prefix is connected with suffixed concrete nouns referring to human actors. For example, *a non-member* is a person who does not belong to a club or party, *a non-resident* is a person who does not reside in the place specified, and *a non-smoker* is someone who does not smoke. Similar words include *non-believer, non-combatant, non-conformist, non-reader, non-specialist, non-striker, non-student, non-subscriber*, etc.

c. 'different from the quality described by the adjectival base'. This nuance appears when the prefix is connected with non-gradable adjectives derived from nouns. In some formations, the prefix is connected with bases denoting quality. For example, *a non-aggressive approach* is an approach that is not aggressive, *a non-economic value* is a value that is not economic, and *non-violent protest* is protest that is not violent. Other words include *non-allergic, non-academic, non-addictive, non-classified, non-contributory, non-essential, non-existent, non-fatal, non-partisan, non-scheduled, non-standard, non-verbal, etc.* In other formations, the prefix is connected with bases denoting nationality. For example, *non-American* is someone who is not American, *non-British* is someone who is not British, and *non-French* is someone who is not French.

Peripherally, the prefix *non-* is connected with other bases to express other meanings. It shows such subtle differences in meaning as:

a. 'devoid of the characteristics described by the nominal base'. In some formations, the privative nuance expressed is impartial in tone. It arises when the prefix is connected with simple abstract nouns denoting non-action. For example, *a non-answer* is an answer that is devoid of adequacy, *a non-problem* is a problem that is devoid of difficulty, and *a non-suit* is a lawsuit that is devoid of evidence. Other words include *non-fiction, non-issue, non-power, non-profit, non-target, non-thought, non-title, non-use*, etc. In other formations, the privative nuance expressed is critical in tone. It arises when the prefix is connected with simple concrete nouns referring to non-humans. For example, *a non-book* is a book that is devoid of value, *a non-entity* is an entity that is devoid of consequence, and *a non-event* is an event that is devoid of excitement. In a few formations, the nouns refer to humans. For example, *a non-descript* is a person who is devoid of distinctiveness, and *a non-person* is a person who is devoid of importance.

b. 'resisting the action described by the verbal base'. This nuance appears when the prefix is connected with verbs.[8] Words formed in this way usually belong to commercial jargon. For example, *a non-iron suit* is a suit that does not require ironing, *a non-skid surface* is a surface that resists skidding, *a non-slip mat* is a mat that resists slipping, and *a non-stick pan* is a pan that resists sticking.[9]

Figure 11 presents a graphical representation which captures the multiple uses of the negative prefix *non-*. Note that the solid arrow represents the prototypical sense, whereas the dashed arrows represent the semantic extensions.

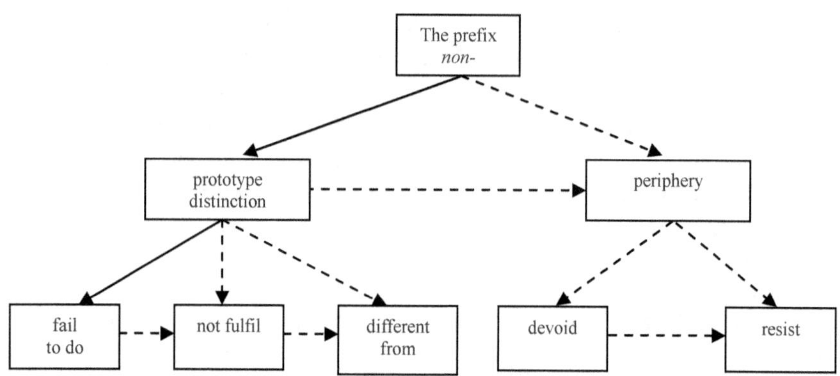

Figure 11: The semantic network of the negative prefix *non-*

3.4.1.6 *un-*

The negative prefix *un-* is from Old English. It is the native English prefix for negation, but it combines freely with non-native bases as well. It is the most common negative prefix in English. It is tied to bases of different lexical classes. In all the lexical formations, the prefix *un-* is class-preserving.

Prototypically, the prefix *un-* is tied to adjectival bases to express distinction, having the following semantic variants.

a. 'the antithesis of what is specified by the adjectival base'. This meaning proceeds when the prefix is tied to adjectives, simple or complex, describing humans. The formations are gradable and express contrariety. In some cases, it describes traits. For example, *uncooperative* is the antithesis of cooperative, *unfair* is the antithesis of fair, and *ungrateful* is the antithesis of grateful. A handful of other words includes *unaccustomed, unaware, unbiased, unfaithful, unhelpful, unkind, unlucky, unwise*, etc. In some cases, it describes nationality. For example, *unAmerican* means the antithesis of an American national in characteristics, *unBritish* means the antithesis of a British national in characteristics, and *unFrench* means the antithesis of a French national in characteristics.

b. 'distinct from what is specified by adjectival base'. This meaning proceeds when the prefix is tied to adjectives, simple or complex, describing non-humans. The formations are gradable and express contrariety. For example, *unofficial* is distinct from being official, *unremarkable* is distinct from being remarkable, and *unsafe* is distinct from being safe. A handful of other words includes *unbelievable, unclean, unclear, uncommon, unintelligible, unnecessary, untidy, unusual*, etc.

c. 'not subjected to what is specified by adjectival base'. This meaning proceeds when the prefix is tied to complex adjectives denoting quality, participles ending in *-ed* or *-ing*. For example, *undressed* means not dressed, *uneducated* means not educated, and *unfinished* means not finished. A handful of other words includes *unaltered, unbaked, unchanging, undemanding, unexamined, uninterrupted, uninviting, unoffending, unsmiling, untouched*, etc.

Peripherally, the prefix *un-* is tied to other bases to express other meanings, having the following semantic variants:

a. 'inverting what is specified by the verbal bases'. The meaning of reversal proceeds when the prefix is tied to verbal bases denoting action, meaning that the object has the physical ability to undergo change. For example, *unclose* means inverting the action of closing, *unpack* means inverting the action of packing, and *unsettle* means inverting the action of settling. A handful of other words includes *undo, unclench, unfreeze, unlearn, unlive, unloose, unmake, unscrew, unteach, unthink,* etc. In some cases, the bases can be both nominal and verbal. For example, *unclip* means inverting the action of clipping, *unlace* means inverting the action of lacing, and *unzip* means inverting the action of zipping. A handful of other words includes *unbutton, unclamp, uncover, unleash, unlock, unplug, unroll, unseal, untie, unwrap*, etc.[10]

b. 'taking away what is specified by the nominal base'. The meaning of removal proceeds when the prefix is tied to concrete nouns, with the resulting formation denoting separation or release. For example, *unchain* means taking away a chain from somebody or something, *unhook* means taking away a hook from something, and *unload* means taking away a load from a ship or vehicle. A handful of other words includes *unbrace, uncage, uncurl, unfrock, unhand, unharness, unhorse, unman, unmask, unseat,* etc.

c. 'bereft of what is specified by the nominal base'. The meaning of privation proceeds when the negative prefix is tied to abstract nouns. In some formations, the nouns imply non-action. For example, *unease* means something which is bereft of ease, *unrest* means a state which is bereft of rest, and *untruth* means something which is bereft of truth. A handful of other words includes *unculture, unhealth, unintelligence, unlaw, unpeace, unscience, unwisdom*, etc. In other formations, the nouns imply action. For example, *unbelief* means lacking belief, *unconcern* means lacking worry, and *unsuccess* means lacking success.[11]

Figure 12 presents a graphical representation which captures the multiple uses of the negative prefix *un-*. Note that the solid arrow represents the prototypical sense, whereas the dashed arrows represent the semantic extensions.

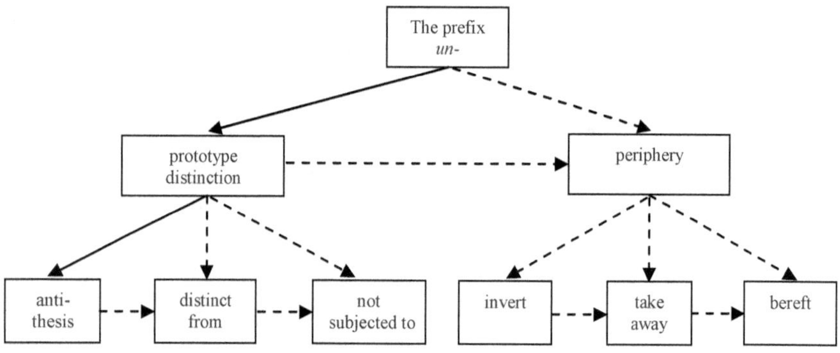

Figure 12: The semantic network of the negative prefix *un-*

3.4.2 Secondary negative prefixes

Secondary negative prefixes are bound morphemes which indicate the adverse of the base to which they are attached. They are considered secondary because they are not negative per se, but connote negation. In addition to *ab-* and *semi-* which were discussed in the previous chapters, secondary negative prefixes include *anti-, contra-, counter-, mal-, mis-, pseudo-, quasi-, sub-* and *under-*. Below is a prototype-based characterisation of each of them.[12]

3.4.2.1 *anti-*

Prototypically, the prefix *anti-*, which originates in Greek, is appended to nominal bases to express opposition. In this use, it is class-preserving although in some example it could be deemed class-changing. Relying on the nature of the integrated base, the prefix *anti-* has the following senses.[13]

a. 'reacting against the thing named by the nominal base'. This meaning emerges when the prefix is appended to complex abstract nouns. In this use, the prefix functions to describe a physical action performed by humans against certain practices. For example, *an anti-abortion campaign* is a campaign that reacts against abortion, *an anti-discrimination slogan* is a slogan that reacts against discrimination, and *anti-war sentiment* is sentiment that reacts against war. More words exemplifying this use are *anti-apartheid, anti-colonialism, anti-establishment, anti-immigration, anti-slavery,* etc.

b. 'opposed to the thing named by the base'. This meaning emerges when the prefix is appended to complex nouns or adjectives ending mostly in *-ist*. In this use, the prefix functions to describe a verbal action performed by humans against certain concepts. For example, *an anti-communist demonstrator* is a demonstrator who is opposed to the concept of communism, *an anti-federalist group* is a group that is opposed to the concept of regional government, and *an anti-racist alliance* is an alliance that is opposed to the concept of racism. More words exemplifying this use are *anti-capitalist, anti-fascist, anti-imperialist, anti-socialist, anti-terrorist*, etc. In some cases, the prefix precedes names of countries. *An anti-X politician* is a politician who is opposed to the government of the X country and its policies.

c. 'displaying the opposite characteristics of the thing named by the nominal base'. This meaning emerges when the prefix is appended to simple nouns. In this use, the prefix functions to describe an entity which fails to display the essential characteristics of another. For example, *anti-climax* displays the opposite characteristics of a climax like triviality and insignificance, *anti-hero* displays the opposite characteristics of a hero like inaction, indecision and weakness, and *antiparticle* displays the opposite characteristics of a particle vis-à-vis electric charge or magnetic effect.

Peripherally, the prefix *anti-* is appended to nominal bases to express obstruction. It is class-preserving. Relying on the nature of the integrated base, the prefix *anti-* has the following senses:

a. 'preventing the thing named by the nominal base'. This meaning emerges when the prefix is appended to abstract nouns implying non-action and denoting medical substances. In this use, the prefix functions to mean that one substance is being used to prevent the effects of another. For example, *an anti-bacteria chemical* is a chemical used to kill bacteria, *an anti-cancer therapy* is a therapy used to combat cancer, *an anti-cholesterol drug* is a drug used to reduce cholesterol, and *an anti-coagulant substance* is a substance used to prevent blood clotting.

b. 'hindering the action named by the nominal base'. This meaning emerges when the prefix is appended to abstract nouns implying action and denoting commercial means. In this use, the prefix functions to signify that one means is being used to protect one against another. For example, *anti-freeze liquid* is liquid that hinders something from freezing, *anti-inflation measure* is a measure that hinders prices from inflating, and *anti-pollution law* is a law that protects the environment against pollution.

c. 'defending against the weapon named by the nominal base'. This meaning emerges when the prefix is appended to concrete nouns denoting military weaponry. In this use, the prefix functions to mean that one weapon is being used to combat or destroy another. For example, *an anti-aircraft weapon* is a weapon designed to defend against aircraft, *an anti-missile rocket* is a rocket designed to defend against missiles, and *an anti-tank gun* is a gun designed to defend against tanks.

Figure 13 presents a graphical representation which captures the multiple uses of the negative prefix *anti-*. Note that the solid arrow represents the prototypical sense, whereas the dashed arrows represent the semantic extensions.

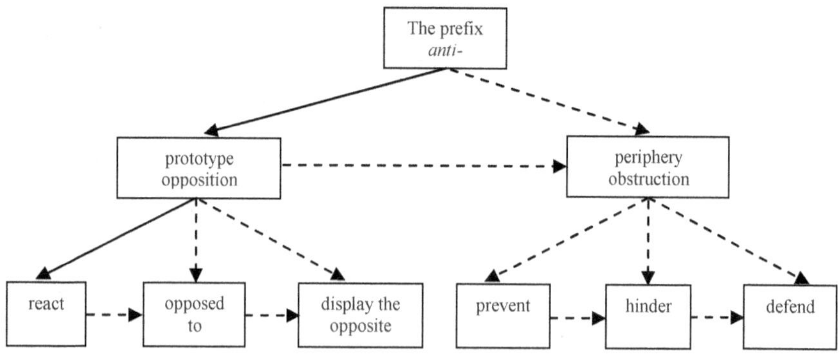

Figure 13: The semantic network of the negative prefix *anti-*

3.4.2.2 *contra-*

Prototypically, the prefix *contra-*, which is borrowed from Latin, is tacked on to nominal bases to express opposition. In this use, it is class-preserving. Relying on the nature of the annexed base, the prefix *contra-* is interpreted as:

a. 'in comparison with the thing mentioned in the nominal base'. This meaning comes to the surface when the prefix is tacked on to abstract nouns implying non-action. For example, *a contra-distinction* is a distinction that is in comparison with another, *a contra-indication* is an indication that is in comparison with another, used as a reason against the use of a particular medicine, and *a contra-position* is a position that is in comparison with another.

b. 'posed against the thing mentioned in the nominal base'. This meaning comes to the surface when the prefix is tacked on to abstract nouns implying action. For example, *a contra-flow* is a flow in a traffic system posed against another, and *a contra-rotation* is a motion in a circle posed against another.

Peripherally, the prefix *contra-* is tacked on to nominal bases to express degree, having the meaning of 'pitched lower or higher than the thing mentioned in the nominal base'. This meaning comes to the surface when the prefix is attached on to concrete nouns denoting musical instruments. For example, *a contrabass* is an instrument pitched an octave below the usual range of that instrument, *a contragamba* is an instrument which is pitched an octave higher than normal, and *a contratenor* is a singer whose voice is lower than that of a tenor.

Figure 14 presents a graphical representation which captures the multiple uses of the negative prefix *contra-*. Note that the solid arrow represents the prototypical sense, whereas the dashed arrows represent the semantic extensions.

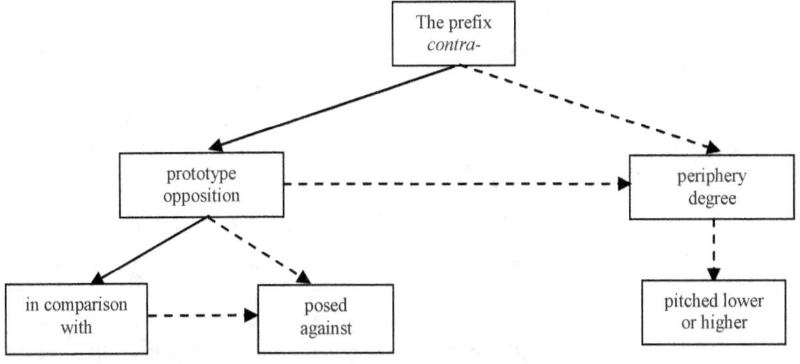

Figure 14: The semantic network of the negative prefix *contra-*

3.4.2.3 counter-

Prototypically, the prefix *counter-*, which comes from French, is adjoined to nominal bases to express opposition. In this use, it is class-preserving. Relative to the nature of the joined base, the prefix *counter-* is analysed as:

a. 'standing against the thing designated by the nominal base'. This meaning happens when the prefix is adjoined to abstract nouns denoting non-action. In this sense, the prefix picks out the meaning that one thing is offered as an alternative to or in return for another. For example, *counter-balance* is balance that stands against another, *a counter-example* is an example that stands against another in an argument, and *a counter-proposal* is a proposal that stands against another in a discussion. More examples of this sense are *counter-culture, counter-evidence, counter-influence, counter-point, counter-pressure, counter-propaganda, counter-strategy, counter-weight*, etc.

b. 'facing the action or activity designated by the nominal base'. This meaning happens when the prefix is adjoined to abstract nouns denoting action. Some of these examples can also be used as verbs. In this sense, the prefix picks out the meaning that one action is taken as a response to or in retaliation against another. For example, *a counter-attack* is an attack that is carried out to face another, *a counter-claim* is a claim set up to face another, and *a counter-move* is a move that is made to face another. More examples of this sense are *counter-blow, counter-demonstration, counter-espionage, counter-march, counter-offensive, counter-punch, counter-strike, counter-terrorism*, etc.

Peripherally, the prefix *counter-* is adjoined to other bases to express correspondence. It is class-preserving. Relative to the nature of the joined base, the prefix is analysed as:

a. 'matching the thing designated by the nominal base'. This meaning happens when the prefix is adjoined to abstract nouns denoting non-action. In this sense, the prefix shows that one entity is similar or equal to another. More precisely, it emphasises the opposite constituent of a thing that has naturally opposite parts. For example, *a counter-part* is a matching part, *a counter-scale* is a matching scale, and *a counter-type* is a matching type. More examples of this sense are *counter-balance, counter-foil, counter-poise, counter-stock, counter-tally, counter-weight*, etc.

b. 'duplicating the thing designated by the nominal base'. This meaning happens when the prefix is adjoined to abstract nouns denoting action. In this sense, the prefix shows that one thing can be used as a substitute for another. For example, *a counter-check* is a double check, as for accuracy, *a counter-signature* is a signature added to a document that has already been signed.

Figure 15 presents a graphical representation which captures the multiple uses of the negative prefix *counter-*. Note that the solid arrow represents the prototypical sense, whereas the dashed arrows represent the semantic extensions.

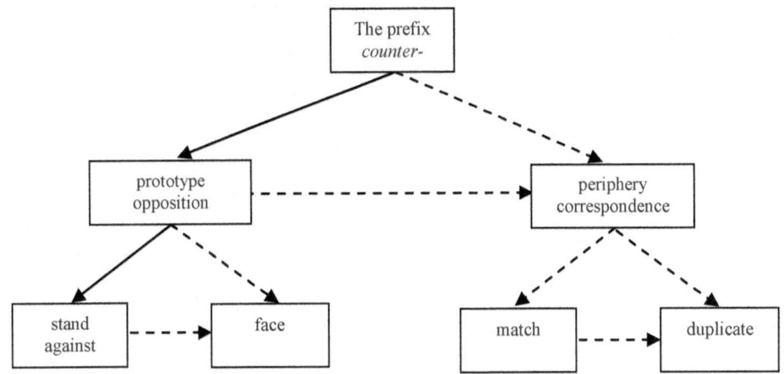

Figure 15: The semantic network of the negative prefix *counter-*

3.4.2.4 mal-

Prototypically, the prefix *mal-*, whose origin is Latin, is bound with verbs and their derivatives to express treatment. In this use, *mal-* is class-preserving, and conveys the following shades of meaning:

a. 'inappropriately executing the action denoted by the verbal base'. This meaning comes to attention when the prefix is bound with verbal bases denoting action. Precisely, it means something done in a faulty or imperfect manner. For example, *maladminister* means to administer badly, inefficiently or dishonestly, *malfunction* means to function incorrectly or irregularly, and *maltreat* means to treat badly, cruelly or inconsiderately.

b. 'inappropriate execution of the thing denoted by the nominal base'. This meaning comes to attention when the prefix is bound with nouns denoting result. For example, *maladjustment* means a bad, defective or imperfect adjustment, *maldistribution* means a faulty, unequal or unfair distribution, and *malpractice* means an immoral, illegal or unethical practice. More coinages include *maladministration, malcontent, malformation, malnutrition, maltreatment*, etc.

Peripherally, the prefix *mal-* is bound with adjectival bases or participles denoting a condition. Its meaning is 'not characterised by the thing denoted by the adjectival base'. The new formations are gradable. For example, *maladroit* means not characterised by tactfulness in behaviour, *malformed* means not characterised by normality in form, *malodorous* means not characterised by pleasantness in smell, and *malnourished* means not characterised by sufficiency in food.

Figure 16 presents a graphical representation which captures the multiple uses of the negative prefix *mal-*. Note that the solid arrow represents the prototypical sense, whereas the dashed arrows represent the semantic extensions.

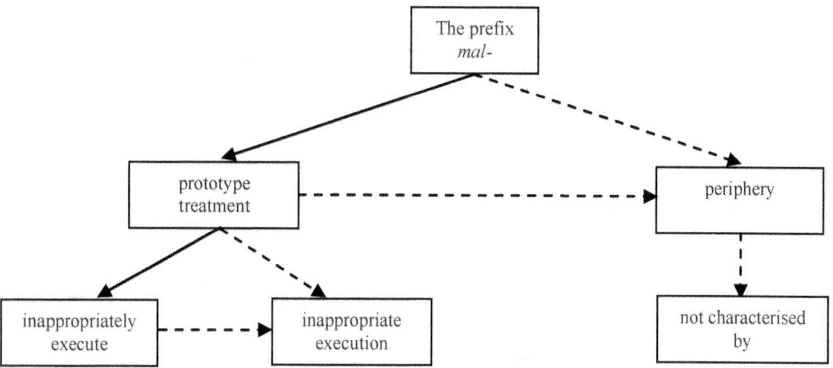

Figure 16: The semantic network of the negative prefix *mal-*

3.4.2.5 *mis-*

Prototypically, the prefix *mis-*, whose source is Old English, is predominantly fused with verbs and their derivatives to express treatment. In this use, *mis-* is class-preserving, and has the following shades of meaning:

a. 'falsely performing the action conveyed by the verbal base'. This meaning comes out when the prefix is fused with verbal bases marking action, mostly denoting mental acts. Precisely, it means doing something in a faulty or imperfect manner. For example, *misconceive* means conceiving in the wrong way, *misrepresent* means representing wrongly or inaccurately, and *misunderstand* means understanding wrongly or incorrectly. Additional examples include *misapprehend, miscalculate, misconstrue, misinterpret, misjudge, misquote, misread, misrule*, etc.

b. 'false performance of the thing conveyed by the verbal base'. This meaning comes out when the prefix is fused with nouns marking result, mostly denoting physical objects. Precisely, it means something done in a defective or flawed manner. For example, *miskick* means to kick a ball wrongly, *mishandle* means to handle something wrongly, and *mismanage* means to manage something badly. Additional examples include *misapply, misappropriate, mischarge, misdirect, mishit, mislay, misplace, misthrow*, etc.[14]

Peripherally, the prefix *mis-* is fused with nouns marking events, mostly denoting behaviour. The prefix means 'improper performance of the thing conveyed by the nominal base'. For example, *misadventure* means a small accident, *mischance* means an unlucky event, and *misdemeanour* means a minor offence. Additional examples include *misbehaviour, miscarriage, misconduct, misfit, misgiving, misshape*, etc.

Figure 17 presents a graphical representation which captures the multiple uses of the negative prefix *mis-*. Note that the solid arrow represents the prototypical sense, whereas the dashed arrows represent the semantic extensions.

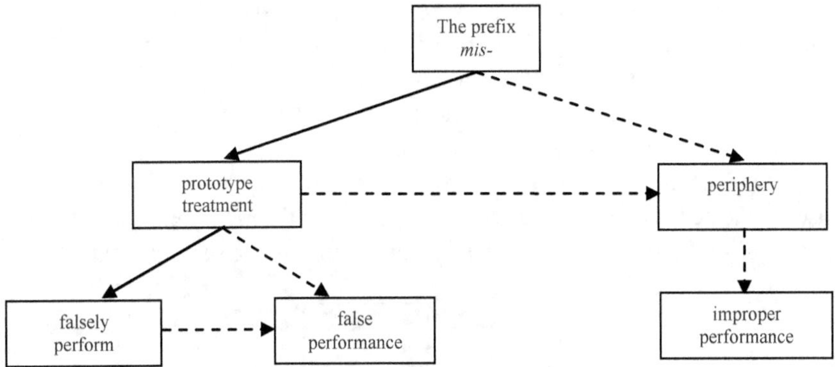

Figure 17: The semantic network of the negative prefix *mis-*

3.4.2.6 *pseudo-*

Prototypically, the prefix *pseudo-*, which comes from Greek, is added to various bases to express inadequacy. In this use, it is class-preserving. Relative to the nature of the joined base, the prefix *pseudo-* means:

a. 'simulating the thing indicated by the nominal base'. In this use, the prefix is added to nouns, usually biological, to refer to organs which have a function other than the one you might expect, or species which resemble another despite being unrelated to it. For example, *a pseudo-bulb* is an enlarged portion of stem located above the ground, as in many tropical orchids, *a pseudo-carp* is a fruit formed by combining the ripened ovary with another structure, often the receptacle, e.g. a strawberry, and *a pseudo-scorpion* is a small non-venomous arachnid resembling a tailless scorpion.

b. 'posing as the thing indicated by the adjectival base'. In this use, the prefix is added to adjectives to denote falseness of appearance, i.e. describe things which are not real, i.e. false. For example, *a pseudo-democratic ritual* is a ritual that appears to be democratic but it is not, *a pseudo-dramatic build-up* is a build-up that purports to be dramatic, and pseudo-natural enclosures are enclosures that appear natural but are not. In some formations, the bases are nominal such as *pseudo-military, pseudo-oak, pseudo-science*, etc.

Peripherally, the prefix *pseudo-* is added to nominal bases to express inadequacy, and consequently disapproval. In this use, it is class-preserving. It means 'pretending to be the thing indicated by the nominal base'. In this use, the prefix describes people who are not really what they claim to be, i.e. they are just a sham. For example, *a pseudo-friend* is one who appears to be friendly but is not a genuine friend, *a pseudo-intellectual* is someone who engages in false intellectualism or is intellectually dishonest, and *a pseudo-scientist* is a practitioner of pseudoscience, a system of thought or a theory which is not formed in a scientific way.

Figure 18 presents a graphical representation which captures the multiple uses of the negative prefix *pseudo-*. Note that the solid arrow represents the prototypical sense, whereas the dashed arrows represent the semantic extensions.

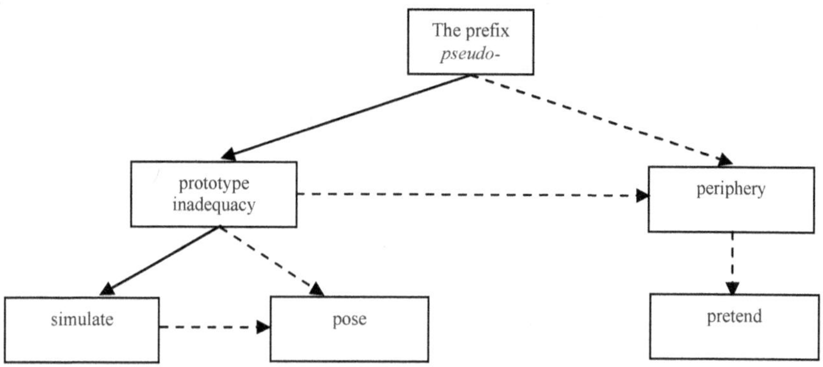

Figure 18: The semantic network of the negative prefix *pseudo-*

3.4.2.7 *quasi-*

Prototypically, the prefix *quasi-*, which comes from Latin, is attached to adjectival bases to express inadequacy. In this use, it is class-preserving and means 'apparently the same as that denoted by the adjectival base'. It suggests that things are not what they seem, i.e. things which appear to be something are actually not. For example, *a quasi-autonomous company* is a company that is apparently independent and has the power to make its own decisions, *a quasi-judicial agency* is an agency that exercises powers or functions like those of a court or a judge, and *a quasi-official organisation* is an organisation which apparently has the approval or authorisation of a public body.

Peripherally, the prefix *quasi-* is attached to adjectival bases to express inadequacy. In this use, it is class-preserving and means 'having some, but not all of the features of the thing denoted by the base'. It suggests that things are to some extent or degree similar to something else, i.e. things which are almost but not quite the same as that of another. For example, *a quasi-independent commission* is a commission that is to a degree, but not completely independent, *quasi-crystal material* is material that is partly crystal, and a *quasi-military uniform* is uniform that is almost but not completely military.

Figure 19 presents a graphical representation which captures the multiple uses of the negative prefix *quasi-*. Note that the solid arrow represents the prototypical sense, whereas the dashed arrows represent the semantic extensions.

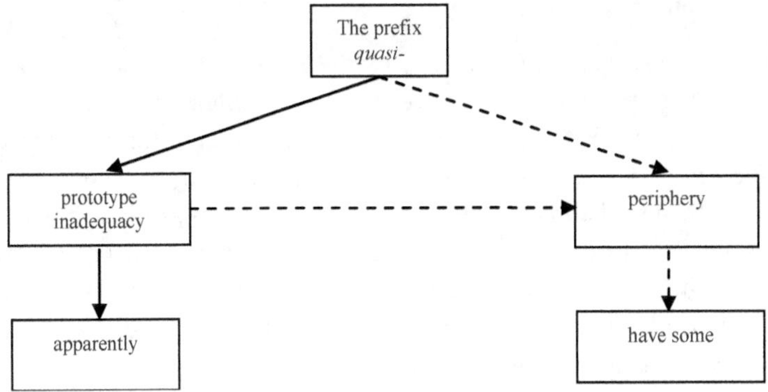

Figure 19: The semantic network of the negative prefix *quasi-*

3.4.2.8 sub-

Prototypically, the prefix *sub-*, which comes from Latin, is appended to various bases to express degradation. In this use, it is class-preserving. Relative to the nature of the joined base, the prefix *sub-* acquires the following senses:

a. 'below or beneath the thing named by the nominal base'. In this use, the prefix is appended to nouns to indicate position, i.e. to show that something is lying underneath or lower than something else. For example, *a submarine* is a ship that can travel under the sea, *subsoil* is the soil lying immediately under the surface soil, and *a subway* is a tunnel under a road for use by pedestrians. Other examples include *sub-basement, sub-current, sub-floor, sub-railway, sub-structure, sub-surface, sub-trench, sub-zero*, etc.

b. 'forming part of the thing named by the nominal base'. In this use, the prefix is appended to nouns to indicate structure, i.e. to show that one thing forms a smaller part of a larger whole. For example, *a sub-class* is a smaller group among several into which a main class is divided, *a sub-district* is a division of a district, and *a sub-family* is a subdivision of a family in the classification of animals, plants and languages. Other examples include *sub-arch, sub-base, sub-branch, sub-committee, sub-continent, sub-contract, sub-distinction, sub-group, sub-section, sub-title*, etc. In some formations, the bases are verbal such as *sub-classify, sub-colonise, sub-divide, sub-lease, sub-let*, etc.

c. 'almost or nearly the thing named by the adjectival base'. In this use, the prefix is appended to adjectives to describe inferiority, i.e. to show that things are less complete or smaller than others. For example, *a sub-clinical infection* is an infection whose symptoms are not detectable by the usual clinical tests, a *sub-sonic aircraft* is an aircraft which flies at a speed slower than that of sound, and *sub-standard goods* are goods which fail to meet a required standard. Other examples include *sub-antarctic, sub-fertile, sub-freezing, sub-humid, sub-literate sub-normal, sub-tropical*, etc.

Peripherally, the prefix *sub-* is appended to nominal bases to express degradation. In this use, it is class-preserving. It means 'subordinate to the thing named by the nominal base'. In this use, the prefix is attached to personal nouns to indicate rank, i.e. to show that someone is secondary or inferior to another. For example, *a sub-dean* is the deputy or substitute of a dean, *a sub-editor* is an assistant editor helping to prepare material for publication, and *a sub-governor* is a subordinate or assistant governor. Other examples include *sub-agent, sub-commissioner, sub-head, sub-lieutenant, sub-officer, sub-treasurer, sub-warden*, etc.

Figure 20 presents a graphical representation which captures the multiple uses of the negative prefix *sub-*. Note that the solid arrow represents the prototypical sense, whereas the dashed arrows represent the semantic extensions.

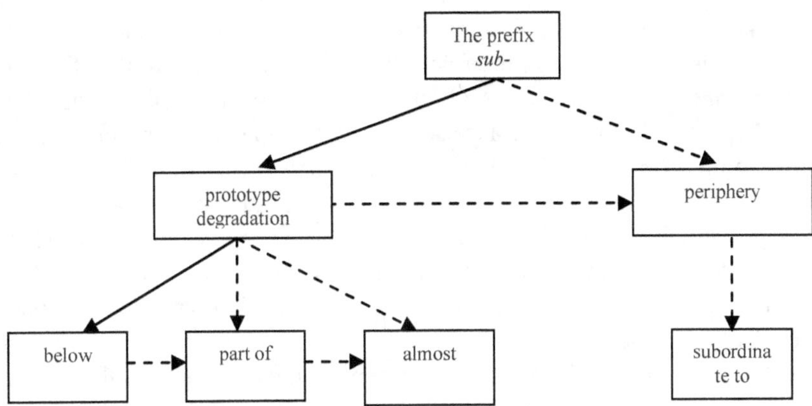

Figure 20: The semantic network of the negative prefix *sub-*

3.4.2.9 *under-*

Prototypically, the prefix *under-*, which comes from Old English, is annexed to various bases to express degradation. In this use, it is class-preserving. Relative to the nature of the joined base, the prefix *under-* has the following senses:

a. 'below or underneath the thing expressed by the nominal base'. In this use, the prefix is annexed to nouns to describe the position of things. For example, *the underground* is a railway system in which electric trains travel along passages below the surface of the ground, *an underpass* is a road or path that goes under something such as a busy road, allowing vehicles or people to go from one side to the other, and *underwear* is clothing worn under other clothes next to the skin. Other examples include *underarm, undercarriage, underclothes, undercover, undercurrent, underfoot, underslip, under*tow, etc.

b. 'less in degree or quantity than the thing expressed by the adjectival base'. In this use, the prefix is annexed to adjectives, usually past participles, to describe insufficiency, i.e. to show that things are less in degree, rate, or quantity than normal. For example, *undercooked potatoes* are

> potatoes which are not cooked enough, *undernourished children* are children who do not eat enough to maintain good health, and *understaffed schools* are schools which do not have enough employees. Other examples include *under-developed, under-equipped, under-financed, under-manned, under-privileged, under-sized, under-trained*, etc. In some formations, the bases are verbal such as *under-estimate, under-grow, under-pay, under-state, under-use, under-value*, etc.

Peripherally, the prefix *under-* is annexed to nominal bases to express degradation. In this use, it is class-preserving. It means 'lower in status than the thing expressed by the nominal base'. In this use, the prefix describes people who have a lower rank than someone else, or are less important than someone else. For example, *an underachiever* is a person who performs less well or achieves less success than expected, *an undergraduate* is a student at a university who has not yet taken a first degree, and *an undersecretary* is a person who has a slightly lower rank than a secretary. The same meaning is true when the prefix is annexed to numbers. For example *under-fives* are children who are younger than five years old.

Figure 21 presents a graphical representation which captures the multiple uses of the negative prefix *under-*. Note that the solid arrow represents the prototypical sense, whereas the dashed arrows represent the semantic extensions.

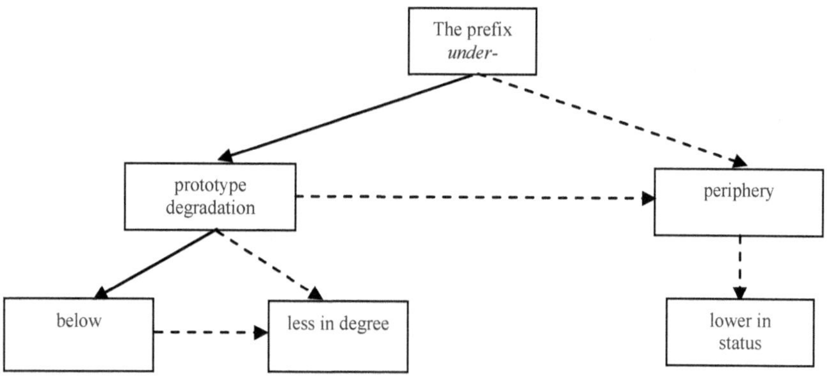

Figure 21: The semantic network of the negative prefix *under-*

3.5 Summary

In this chapter, I applied the *prototype* model to the description of negative prefixes. The model is based on the notion of polysemy. As Fillmore (1982) states, polysemy is the range of meanings a lexeme has as a result of the process of meaning extensions. Any negative prefix forms a category of distinct but related senses. It has a central sense with all the other senses linked to it along various dimensions. A category of any negative prefix has a sense at the centre which is extended to form derived senses. The meaning extensions are not governed from the centre by rules, but rather by convention. Meanings are not the result of primitive concepts, but rather of psychologically-motivated concepts. Meaning extensions are explainable in terms of motivation, not features, as has been posited previously. As Taylor (1989) claims, the meanings of any lexeme are chained in such a way that everything is associated with everything else. Unlike the classical view which deems lexical items to be static in nature, the cognitive view considers them dynamic in nature. The meanings associated with lexical items are not absolute or fixed. Rather, they are flexible and capable of extending over time and through use.

Casting a quick look at the data discussed so far leads us to a number of conclusions regarding morphology:

1) The interpretation of a negative word follows from the property of the prefix together with the semantic nature of the base. Polysemy arises from the interaction between the meanings attributed to the prefix with the meanings of the different kinds of bases. The negative prefixes of English are therefore a good example of polysemy.

2) Negative prefixes come from different sources. The prefixes *a-*, *anti-* and *pseudo-* are of Greek origin. The prefixes *ab-*, *contra-*, *de-*, *dis-*, *in-*, *mal- non-*, *quasi-*, *semi-* and *sub-* come from Latin. The prefix *counter-* is of French origin. The prefixes *mis-*, *under-* and *un-* are the only prefixes that are from Old English.

3) Negative prefixes differ with reference to their ability to change the classes of the bases to which they are added. The prefixes *a(n)-*, *anti-*, *contra-*, *counter-*, *in-*, *mal-*, *mis-*, *non-* and *un-* are class-preserving. Prototypically, the prefixes *de-* is class-changing. In its prototypical use, the prefix *dis-* is class-preserving.

4) Negative prefixes attach to all kinds of bases. Prototypically, the prefixes *a(n)*, *dis-*, *in-* and *un-* attach to adjectival bases. Prototypically, the prefixes *contra-*, *counter-* and *non-* attach to nominal bases. Peripherally, these prefixes attach to other classes of bases. The prefixes *anti-* and *de-* attach only to nominal bases.

5) Negative prefixes attach to both gradable and non-gradable bases, and quite consistently form negatives that are both contradictory and contrary in meaning. The contrary reading appears when the bases

have gradable interpretations. The contradictory reading arises from bases that have non-gradable interpretations. Among the prefixes, only *non-* attaches to non-gradable bases.

Taking a close look at the data discussed so far leads us to a number of conclusions regarding semantics.

1) Negative prefixes qualify as dependent (bound morphemes), whereas bases qualify as autonomous (free morphemes). The prefix *ab-*, for example, cannot exist on its own. To complete its meaning, it needs a base like *normal* to form *abnormal*. Though dependent, negative prefixes are responsible for the character of the resulting formations. The prefix *mis-*, for example, which means 'bad or wrong', lends its profile to the word *misbehaviour* to mean 'bad behaviour'.

2) No single definition accounts for all the occurrences of a negative prefix. Rather, a small number of minimally-distinct definitions account for the various uses as well as for the differences in distribution. The distinction between the different senses of a negative prefix is a consequence of the different environments in which it occurs.

3) The polysemy of a negative prefix is not random or idiosyncratic. Rather, it is structured in that there are certain principles which account for the sense extensions and the relationship among them. In other words, one can specify the respects in which the senses are distinguishable by appealing to semantic principles like *animacy, concreteness, property* and *transitivity*, which are systematic across a segment of the lexicon.

4) Negative prefixes are versatile in that they can be used for many different purposes. The difference between one prefix and another is a consequence of semantics, namely of the different concepts which they represent. For example, the prefixes *de-, dis-* and *un-* symbolise *privation, removal* and *reversal*. The differences reside in the lexical semantics of the particular sets of bases which they favour.

5) Negative prefixes can, in rare cases, function as intensifiers. For instance, in *denude*, the negative prefix *de-* means 'emphatically', as in *The hillsides are denuded of trees*. In *unloose*, the negative prefix *un-* means 'completely', as in *He unloosed his tie*.

6) Negative prefixes allow either a contrary or contradictory reading, which is generally correlated with the nature of the base. Gradable bases yield contrary readings, where an intermediate state is percep-

tible. Non-gradable bases allow contradictory readings, where an intermediate state is not perceptible. A contrary reading carries an evaluative force, whereas a contradictory reading carries a descriptive force. Of all the negative prefixes, *non-* and *a(n)*, to a lesser degree, favour a contradictory reading.

Notes

1 The negative prefix *a-* should not be confused with the letter *a* in *ago, asleep* or *aside*, which does not have anything to do with negation. Nor should it be confused with the letter *a* found mostly in scientific terminology as in *apnea, anaemia* or *apraxia*.

2 Andrews (1986: 223) describes the resulting verbs derived by the negative prefix *de-* as effective in nature, i.e. they create the object as a direct result of the verbal process.

3 Andrews (1986: 223) describes the resulting verbs derived by the negative prefix *de-* as affective in nature, i.e. they do not create the object. Rather, the object exists prior to the verbal process.

4 Lieber (2004: 116) states that the negative prefixes *dis-* and *un-* do not occur with verbal bases denoting manner, that is, verbs which imply some change, but not a directed change, as in **disrun*, **diswalk*, **disvary*, **unrun*, **unwalk*, **unvary*, etc.

5 The negative prefix *in-* should not be confused with *in-* in words like *inborn, implant* or *inhale*. In these words, the *in-* is not a marker of negation. Of particular interest is the word *inflammable* which means the same thing as *flammable*. In modern English, the word, as Quirk et al, 1985: 1541) write, is often replaced by *flammable* with its opposite *non-flammable*.

6 As Quirk et al. (1985: 1540) point out, some negative lexicalisations with *in-* and its variants have no direct relation to the non-negative bases. For example, the word *infamous* does not mean 'not famous', but 'well known for being bad or evil'. The word *infirm* does not mean 'not firm', but 'sick and weak, especially over a long period or because of old age'.

7 In British usage, *non-* is joined to a word by means of a hyphen. In American usage, *non-* is joined without a hyphen, but a hyphen is preferred if its absence would obscure the understanding or pronunciation of the word. Some words prefixed by *non-* rarely or never use a hyphen such as *nonentity, nonsense*, etc.

8 In Lieber's (2004: 114) view, the negative prefix *non-* does not attach to verbs and therefore does not display either a reversal or a privative meaning.

9 For the voguish uses of *non-* and a glossary of the words that host it in English, especially American English, the reader is referred to Algeo (1971: 87–105).

10 According to Lieber (2004: 117), the negative prefix *un-* cannot attach to a causative or inchoative verb like *explode* (*unexplode), because its resulting state is quite permanent. Besides, whenever there is a common word which is the

opposite, the *un-* form does not exist, as in *high-low* vs. **unhigh-*unlow*, and *fast-slow* vs. **unfast-*unslow*.

11 The prefix *un-* can be used with *learn* and *say* to denote activity. Horn (2002: 14–5) and Lieber (2004: 116) give insightful analyses of such cases. The words *unlearn* and *unsay* are most comfortably used in negative or subjunctive contexts, as in *You can't unsay what you said* vs. ?*You can unsay what you said*, or *I wish I could unlearn that bad habit* vs. ? *I unlearned that bad habit*.

12 *With-* can also serve as a negative formative in a few formations. Prototypically, it combines with verbs to express *removal* meaning 'away'. For example, *withdraw* means 'to remove something'. Extended from this is a formation in which it expresses *opposition* meaning 'against'. For example, *withstand* means 'to stand up to somebody or something'. Further is a formation in which it expresses *rejection* meaning 'refuse'. For example, *withhold* means 'to refuse to do or give something until something else is done'. Finally, it combines with prepositions to express *privation* meaning 'lack'. For example, *without* means 'lacking the thing indicated by the base'.

13 The negative prefix *anti-* should not be confused with *ante-*, which is a prefix of Latin origin meaning *before* or *in front of*, as in *ante-room, antenatal* or *antedate*.

14 In some formations, *ill-* can also serve as a negative prefix. Prototypically, it combines with past participles to describe an action as being 'badly done'. For example, *an ill-timed comment* is a comment that is made at the wrong time. In some cases, it combines with present participles, as in *ill-fitting clothes*. In other cases, it combines with verbs to describe an action as being 'violent', as in *ill-treat children*. Peripherally, it combines with nouns and adjectives to describe something as being 'unpleasant'. For example, *ill will* means bad feelings between people because of things that happened in the past.

4 Domain

4.0 Overview

This chapter probes the second axis of my semantic system, i.e. the role of domains in the semantic description of negative prefixes as groups of lexemes. To do this, I substantiate two tenets of Cognitive Semantics. One tenet is that the meaning of an expression is described on the basis of the domain it evokes. Applying this tenet to morphology, I argue that the meaning of a prefix can best be defined by linking it to the appropriate facet in the domain in which is located. The second tenet is that the semantic structure of a lexeme relates to conceptual knowledge which is based on human experience. Applying this tenet to morphology, I argue that the use of a prefix is a response to the needs of communication. To that end, the chapter is organised as follows. Section 1 tackles the issue of lexical meaning and surveys the questions raised in its investigation. Section 2 tests out two theories of lexical relationships. One is the dictionary theory, which consists of Semantic Field and Componential Analysis. The other is the encyclopaedic theory, which consists of Frame Semantics and Cognitive Domain. Section 3 applies the assumptions of Cognitive Domain to the description of negative prefixes. Section 4 elaborates on the cognitive domains which the negative prefixes evoke in English. Section 5 provides a summary of the chapter and comes to a number of conclusions.

4.1 Introduction

Semantics is an important sub-discipline within linguistics, and has two concerns. One concern is the meaning of a lexeme. Meaning is the characteristic of a linguistic form which serves to pick out some aspects of the non-linguistic world. Meaning stands at the very centre of the linguistic quest to understand the nature of language. This is because meaning is what languages are all about. Everything in language, words, grammatical constructions, intonation and the like, conspires to realise the goal of meaning in the fullest way. Another concern is the existence of two or more lexemes which may have some semantic relationship. In the literature, there have been many attempts to investigate meaning and explore the relationship among lexemes. While there is a sort of agreement on the issue of meaning, there is a heated controversy over its analysis. In general, the controversy concerns the way in which the lexicon is organised. In particular, the controversy concerns the principles that govern the relationships among the existing lexemes. In this regard, linguists belonging to different linguistic persuasions offer diverse solutions.

In tackling the issue of meaning, two central questions are posed. The first is: if a lexeme has meaning, how can this be identified? The second is: if two or more lexemes are related, how can one tell the difference? The present study attempts to find answers to such questions. The goal is to provide an account of how lexemes in natural language behave and how they are organised. To do this, the chapter is organised as follows. Section 2 reviews the literature which includes the two main theories of meaning. The dictionary theory represents a model which focuses only on linguistic meaning. The encyclopaedic theory represents a model which focuses on both linguistic and non-linguistic meanings. Section 3 forms the nucleus of the study. It extends the domain theory to the analysis of negative prefixes in English. The goal is to show how influential the domain theory is compared with the other theories. Before embarking on the analysis, it is necessary to consider theories of lexical relations. Section 4 introduces the main cognitive domains which the negative prefixes evoke for their characterisation.

4.2 Theories of lexical relationships

A growing area of interest in linguistics has been lexicology, the study of the lexemes of a language. Two aspects are central to the study of lexis. One aspect pertains to the meanings of lexemes and their representation in the lexicon. Every lexeme has a semantic structure of its own. The aim of lexicology is to provide the relevant information about any lexeme, which helps to define its behaviour in language. The other aspect pertains to the relationships between the meanings of lexemes. In this respect, two pieces of information are important. One is paradigmatic information, that is the use of different lexemes for the same concept. The other is syntagmatic information, that is the relation of lexemes to one another when they form a unit. The information provided will ultimately help in understanding language (decoding) and producing language (encoding). Of the two aspects, the present chapter is concerned with the exploration of relationships holding between the meanings of lexical items because they constitute an integral part of word meaning.

To address the issue of lexical relationships, linguists have put forward two general theories of word meaning. One is the dictionary theory, which takes only linguistic phenomena into account. According to this theory, the core meaning of a word, which is stored in our minds, is the information contained in the word's definition. This falls within the sphere of lexical semantics, where the linguist is solely concerned with the non-contextual meaning of the word. This theory is represented by the Semantic Field theory, in which the meaning of a lexeme is described relative to the relationship it holds with its counterparts, and the Componential Analysis theory, according to which the meaning of a lexeme is seen in terms of a number of distinct features. The other general theory of word meaning is the encyclopaedic theory, which takes a range of phenomena, linguistic and non-linguistic, into account. This falls within the sphere of lexical semantics and pragmatics, where the linguist is concerned with the internal and external worlds in shaping linguistic meaning. This theory is represented by the Frame Semantics theory, according to which the meaning of a lexeme is understood by accessing the situation in which it is used, and the Cognitive Domain theory, whereby the meaning of a lexeme is analysed in terms of the conceptual context in which it is embedded. In what follows, I give an overview of the two theories of word meaning together with their affiliates. For more details, see Evans & Green (2006: 215–222).

4.2.1 The dictionary theory

The dictionary theory, which represents only knowledge of linguistic meaning, is based on a number of premises. First, there is a distinction between the core meaning and the pragmatic meaning of a word. According to Evans & Green (2006: 216), *coded* meaning is 'the stored mental representation of a lexical concept'. In Cognitive Semantics, it is called a *schema*, 'a skeletal representation of meaning abstracted from recurrent experience of language use'. From this premise, it follows that meaning is divorced from language use. Coded meaning is the real meaning of a word. Second, all aspects of meaning accessible in a given word are equal. The aspects are structured in terms of positive or negative values. Third, core meaning (semantics) is separated from non-core meaning (pragmatics). Word meaning is a function of semantics. Fourth, words represent fully-specified, pre-assembled meanings. Fifth, the knowledge that a word provides access to is stable. Two theories, adopted within Formal Semantics, represent the dictionary theory. One is the theory of Semantic Field developed by Trier (1931). The other is the theory of Componential Analysis developed by Katz & Fodor (1963).

4.2.1.1 Semantic Field

Semantic Field, also called 'lexical field', is a theory of meaning which goes back to Trier in the 1930s. Basic to the theory is the view that the vocabulary of a language forms clusters of interrelated meanings rather than inventories of independent items. The vocabulary of a language is not simply a listing of independent items, as the head lexemes in a dictionary suggest. Rather, it is organised into fields within which lexemes interrelate and define each other in various ways. The precise meaning of any lexeme can be understood only by placing it in relation to the others. This theory is based on the following premises, which are cited in Bussman (1996: 275). First, the meaning of an individual lexeme is dependent upon the meaning of the rest of the lexemes in the same lexical field. Second, a lexical field is constructed like a mosaic with no gaps; the whole set of all lexical fields of a language reflects a self-contained picture of reality. Third, if a single lexeme undergoes a change in meaning, then the whole structure of the lexical field changes. Examples of semantic fields include *vehicles, fruit, clothing, plants, vision*, and so on. For example, the field of *vision* is divided up into *gape, gawk, gaze, peer, stare*, to name just a few.

The Semantic Field theory has frequently been used by linguists to describe groups of related lexemes, illustrate language change and conduct contrastive analysis of different languages. Yet, as Lehrer (1974) and Lyons (1977) claim, it has some weaknesses. First, it does not sharply differentiate lexemes from one another within the lexical fields. Accordingly, it is not known what position a particular lexeme occupies in the lexical field. Second, not all lexemes are amenable to semantic field analysis. That is, not all aspects of experience neatly divide up into semantic fields. Third, the theory concentrates solely on paradigmatic relations; it does not take into account the contribution made to meaning by syntagmatic information. For example, one cannot say much about the meaning of *bark* without reference to *dog*. Fourth, it does not recognise the role of context in assigning a lexical item to a field. For example, *bank* relates to both the semantic field of institution and that of ground. Fifth, it does not put forward any criterion for determining whether or not a particular lexeme belongs to a lexical field. For more on this, see Lehr (1974, 1978), Lehrer & Kittay (1992) and Lyons (1977, 1995).

4.2.1.2 Componential Analysis

Componential Analysis, also called 'semantic feature analysis', is a theory of meaning which goes back to American anthropologists in the 1950s. Central to the theory is the assumption that the meanings of lexemes can be described in terms of semantic features or primitives. A lexeme is broken down into a list of features which determine its meaning. All lexemes can be decomposed by using a universal, finite set of semantic features. It is on the basis of these features that we organise our experiential world. It is possible to describe the whole lexicon of a language with a limited inventory of universally valid features. Such an analysis allows us to describe more precisely the core meanings of lexemes, which are basic to their individual identity. As a consequence, the analysis enables one to define sense relations among lexemes more closely. Lexemes can be analysed in terms of their semantic components, which usually come in pairs called semantic oppositions. As an illustration of a componential analysis, the lexemes *man/woman* share the features [+adult] and [+human], but differ in terms of the features [+male] vs. [–male] or [–female] vs. [+female]. So, they are incompatible because of gender.

The Componential Analysis theory is still usable in semantic descriptions. However, it has its shortcomings, as described by Bolinger (1965, 1977) and cited in Bussman (1996: 89). First, the features do not correspond directly to the physical properties of the real world. Rather, they are abstract constructs. Second, the features are not yet sufficient to distinguish lexical items. That is, more complex ways must be developed to pin down the meanings of lexemes in areas such as colour, kinship, transitive verbs, etc. Third, the features cannot cope with metaphor or with the fact that much of natural-language meaning resides not only in lexemes but in frozen forms. Fourth, part of the vocabulary can be described through unstructured bundles of semantic features, as in the case of kinship relationships. Fifth, the justification for selecting one value rather than another for a possible component remains unaccounted for. For more on this, see Leech (1981), Katz & Fodor (1963) and Jackendoff (1983, 1990).

4.2.2 The encyclopaedic theory

The *encyclopaedic* theory, which represents knowledge of both linguistic and non-linguistic meaning, makes a number of claims. First, it draws no distinction between semantic and pragmatic knowledge. The meaning of a word subsumes knowledge both of what it means and how it is used. From this premise, it follows that pragmatic meaning is the real meaning of a word. Meaning is a consequence of language use. Second, encyclopaedic knowledge is organised as a network. Aspects of meaning associated with a given word are not equal. Some aspects are more central than others to the meaning of a word. Third, meaning arises from discourse context. The context of use contributes to the encyclopaedic information that a word evokes. Word meaning is a function of context. Fourth, words do not represent neatly pre-packaged bundles of meaning. Instead, they provide access to vast repositories of knowledge relating to a particular concept. Fifth, the encyclopaedic knowledge that a word provides access to is dynamic. New experiences always increase one's knowledge about a given word. Two theories, adopted within Cognitive Semantics, represent the encyclopaedic theory. One is the theory of Frame Semantics developed by Fillmore (1977, 1982), and the other is the theory of Cognitive Domain developed by Langacker (1987: 147–182).

4.2.2.1 Frame Semantics

Frame Semantics is a theory of meaning which was initiated by Fillmore in the 1980s. Fundamental to the theory is the assumption that the meaning associated with a particular lexeme cannot be understood independently of the frame with which it is associated. A frame represents a knowledge structure which relates the lexemes linked with a particular scene. Frames are based on recurring human experience, and have twofold import. They allow us to understand, for example, a group of related lexemes, playing a role in explicating their usage in language. An example that is often cited is that of the commercial transaction frame. One would not be able to understand the lexeme *sell* without knowing anything about the situation of commercial transfer, which involves, among other things, a seller, a buyer, goods, money and so on. For example *sell* views the situation from the perspective of the seller and *buy* from the perspective of the buyer. For more on this, see Fillmore (1977, 1982) and Fillmore & Atkins (1992, 2002).

4.2.2.2 Cognitive Domain

The Cognitive Semantic theory is a similar but not identical theory of meaning. This theory of meaning was formulated by Langacker in the 1980s. Like Fillmore's theory of Frame Semantics, Langacker's theory of Cognitive Domain is based on the assumption that meaning is encyclopaedic in that it covers a large range of knowledge, often in great detail. Lexical concepts cannot be understood independently of the cognitive domains in which they are embedded. A Cognitive Domain is a knowledge structure of varying levels of complexity and organisation. It provides background information against which lexemes can be understood and used in language. Domains are mental experiences, representational spaces or conceptual complexes. For instance, lexemes like *avoid, dodge, elude, escape, eschew, evade, forgo* and *shun* designate different lexical concepts in the domain of *Avoidance*. Without understanding the avoidance domain, we would not be able to use these terms. For more on domains, see Langacker (1987, 1991). Langacker's theory of Cognitive Domain complements Fillmore's theory of Frame Semantics in a number of ways. For a detailed coverage, see Evans & Green (2006: 230–2).

Let us now assess the relevance or irrelevance of the theories. The Semantic Field theory does not distinguish lexemes sharply one from another. The Componential Analysis theory does not concern itself with the actual usages of lexemes. The Frame Semantics theory does not explain the criteria for frame membership or the internal structure of the frame. To advance our understanding of language and better capture how lexical meaning is perceived, I opt here for the Cognitive Domain theory, because it assigns a particular role to each lexeme, focuses on its actual usage, and sets the criteria for its membership in the domain. Above all, this theory gives the speaker the chance to describe a situation differently, using each time a different lexeme. In what follows, I conduct a semantic analysis of the negative prefixes in English, describing their usages in detail and pointing out the relationships between them. The aim is to show how instrumental the Cognitive Domain theory is in resolving questions concerning the specific meaning of a prefix, and the way it contrasts with its counterparts. From this, it becomes obvious that different prefixes form a homogeneous group and that their exact uses are nothing but reflections of diverse experiences.

4.3 Negative prefixes

A negative prefix is added to a base to form a new lexeme. In its primary use, a negative prefix is a bound morpheme which is added to a base to indicate oppositeness. In its secondary use, it is a bound morpheme which is added to a base to signal adverseness. It is bound because it cannot stand on its own, but is combined with a free morpheme. A negative prefix is characterised by two aspects in the derivational process, which ultimately motivate its use. First, prototypically, it does not change the class of the root to which it is added. Both *direct* and *indirect* are adjectives. Second, it clearly contributes to the semantic make-up of the lexeme in which it is a part. It adds meaning to the derived outcome, and serves to reflect the speaker's conceptualisation of a situation. The use of the prefix *in-* serves to signal negation. This is in line with the cognitive assumption that the form of a composite lexeme is motivated by its cognitive organisation, which is determined by both its conceptual content and the construal imposed by the speaker on that content. The construal employed to describe a lexeme's content is realised morphologically by the type of prefix chosen. The use of a prefix is therefore governed by meaning. Each prefix has a different function, and so suits a different context.

The notion of Cognitive Domain has been applied to the study of lexemes in general. It has not to date been explicitly exploited in morphology. In this study, I extend the notion to the description of negative prefixes. To do so, I build on two foundation stones. In the first place, I build on Langacker's claim that semantic structures are characterised relative to cognitive domains. A *domain*, as defined by Langacker (1987: 488), is 'A coherent area of conceptualisation relative to which semantic units may be characterised'. A domain refers to a body of knowledge which provides background information for understanding and interpreting a given item. Langacker (1987: 163) views lexemes as providing access points to knowledge configurations, which constitute clusters of related knowledge, perceptual experiences, cultural values and shared beliefs. Any domain subsumes a number of lexemes, which define its concept in prototypical terms and instantiate its facets in detailed ways. Each lexeme highlights a particular facet of the domain.

Secondly, I also build on my own work (Hamawand, 2007). In this work, I argue that a negative prefix does not exist in isolation in the mind of the speaker but forms together with other conceptually-related prefixes a structured set of elements, which have a reciprocal influence. Central to my approach is the idea that one cannot understand the meaning of a prefix without access to all the essential knowledge that relates to it. A prefix evokes a domain of semantic knowledge relating to the specific concept it refers to. A *Cognitive Domain* is a set of lexemes which cover a certain conceptual area and which bear certain specifiable semantic relations to one another. It is a coherent set of concepts, which are related in such a way that without knowledge of the conceptual structure of the domain, one cannot define the meaning of any one of them. Prefixes not only highlight individual concepts, but also specify a certain perspective in which the domain is viewed. Cognitive Domains are helpful in two ways. First, they provide some insights into the properties of the different prefixes that they subsume. Second, they guide the use of one prefix over the other once the morphological distinctions have been identified.

4.3.1 Assumptions

The subject of the analysis is to present a synchronic description of the behaviour of a lexical item, casting light on the relationship that it holds with its counterparts. To carry out the analysis, the study formulates assumptions based on ideas in Cognitive Linguistics:

1) Lexemes do not exist in isolation in the mind of the speaker, but form, together with other conceptually-related lexemes, structured domains. Domains represent human knowledge of a language, which correspond to background beliefs and perceptual experiences. Domains define implicit conceptual fields. Their facets have peculiarities of their own. Each facet represents a particular aspect of experience. Likewise, negative prefixes cannot be isolated from one another. They gather in domains of different kinds. The domains have specific facets of meaning, which are profiled by the member prefixes. A domain comprises different exemplars, with its attributes being adopted by them.

2) The best way to understand the meaning of a lexeme is to study the domain in which it occurs. The understanding of a lexeme is bound up

with the way it is differentiated from the other related lexemes in the domain. The lexemes in a domain are not independent but are often related conceptually and lexically. Likewise, the meaning of a negative prefix is best understood both by the way it stands for the particular facet of the domain and by the way it stands in a relation of contrast or affinity to the other prefixes in the domain. Speakers store the prefixes together and have extensive knowledge about the relations among them. A domain helps to reveal all the semantic differences which are small but important.

4.3.2 Goals

Having advanced assumptions regarding the issue of lexical relationships, this study now needs to state the goals:

1) Showing that domains mirror the reality of the mental lexicon and provide a natural means of organising it. Domains construe reality in myriad ways and conceive of possible worlds that lie beyond it. Because people constantly experience new things in the world, and because the world is constantly changing, domains allow people to symbolise their conceptualisation readily. Domains account for contextual variability in conceptual representations. Domains are realms of knowledge which provide speakers with the flexibility to construe a situation in different ways. Applied to prefixes, a domain serves to show that the use of a particular prefix represents a particular construal of content.

2) Demonstrating that lexical items applicable to conceptual backgrounds are organised within cognitive domains, where the content of the domains supplies the concepts that are labelled by the lexical items. Domains group together conceptually-related words and account for their meanings, which otherwise would not be possible. Domains are spheres of knowledge within which a number of lexical items can be located, and so are tools for comparing one item with another. Applied to prefixes, a domain serves to explicate similarities and differences among its members, and so is a convenient way of coding information about the distributions of the prefixes and the patterns in which they occur.

3) Indicating that domains uncover the interrelation of concepts in the decomposition of the meaning of a lexical item. Domains represent general information about lexical items and give an account of their

behaviour. Domains are partly constitutive of the meaning of a word, and so can be used as a basis for description. Domains serve as a means of encoding information pertinent to our understanding of a lexical item. Applied to prefixes, a domain serves to provide information about the prefixes, and so are mechanisms for defining their meanings. To define a negative prefix, it is necessary to understand the entire domain and see which of its facets or attributes the prefix picks out or stands for.

4.3.3 Procedures

Having now formulated the assumptions and set out the goals, the study adopts a three-step technique:

1) Identifying the domains, explaining their semantic structures, and exploring their properties. This step realises the first objective that domains serve to organise the lexicon. According to Fillmore (1982), such knowledge structures are necessary in describing the semantic contribution of lexical items and grammatical constructions or the process of constructing the interpretation of a text out of the interpretation of its pieces. The domains evoked by the negative prefixes are *distinction, privation, opposition, removal, reversal* and *treatment.*

2) Selecting the lexical items which belong to the domains by relying on their definitions. This step realises the second objective that domains embrace those lexical items which represent their facets. Domains explain and unify meaning at both conceptual and linguistic levels. The former is about mental content, which is within the mind of the speaker. The latter is about lexical content, which is represented by linguistic symbols. Because of polysemy, it is quite natural to find that some negative prefixes appear in more than one domain, each time serving a different purpose.

3) Stating which lexical item represents which property by comparing it with the other lexical items. This step realises the third objective that domains are interpretative devices by which we understand a lexical item's deployment in a given context. To define a prefix, it is best to compare it with the other prefixes that belong to the same domain. In language, however, it is not surprising to find cases of overlap between lexical items. The same is true of certain prefixes whose meanings overlap. The focus, however, is on the general patterns in which the prefixes occur.

4.4 Prefixal domains

One of the phenomena of language, which is based on real world scenarios, pertains to *negation*. In conceptual terms, *negation* is a background of experience whose nuclear meaning refers to an act of contradiction, denial or inversion. It is the act of indicating oppositeness whereby one thing stands in sharp contrast to another, or signalling adverseness whereby one thing acts against another. In logic, *negation* is a statement of refutation or rebuttal, especially making an assertion that a particular proposition is false. It is a declaration that something is not, has not been, or will not be. Like any chunk of knowledge, negation integrates a number of domains, each of which has its own structure, facets and interconnections. In addition to *inadequacy* and *degradation*, which were previously covered, other examples of negative domains include *distinction, opposition, privation, removal, reversal* and *treatment*. Although these domains are subsumed under negation, each has its own conceptual representation. Each domain has various facets, which reflect physical or social experiences. Each facet spells out the meaning of the domain in a different way. Hence, each facet is characterised by a property of its own.

In morphological terms, *negation* is the process of adding a negative prefix to the beginning of a base to invert its meaning. Like lexical items, negative prefixes cluster together in domains. The way they fit into each domain is determined by their meanings or based on their definitional analyses. Each domain is divided into facets, which are founded conceptually on human experience and represented morphologically by different prefixes. Each prefix is assigned a particular mission to achieve within the domain, and defines itself in contrast to the other members of the domain. In each domain, the negative prefixes are said to stand for the same generic meaning but differ in their specific meaning. A semantic description of any negative prefix must first rely on the whole domain and second on the rival members within the domain. Each negative prefix differs in the kind of message it conveys. As Fillmore & Atkins (1992: 76–7) explain, lexical items are not related directly to one another, but only in virtue of their links to the conceptual facets they represent.

4.4.1 The domain of distinction

The domain of *distinction* is a conceptual area showing contrast or dissimilarity between two entities or their properties. It refers to the process of comparing two or more entities in order to show the differences between them. Entities that are in an opposing direction differ along the contradictory versus contrary lines. A contradictory distinction is the relationship when two opposites exist without allowing any alternative to occur in between, that is two words are contradictory if the positive and negative counterparts admit of no middle ground. For example, a person can be either male or female. This type of distinction amounts to mere description, whose aim is to give an account of what something is, nothing more. A contrary distinction is the relationship when there is a third alternative which is similar to the two opposites but distinct from both. Two words are contrary if the positive and negative counterparts have an intermediate grade. For example, something may be neither hot nor cold; it may be warm. This type of distinction is associated with evaluation, examining something to judge its value, quality or importance.

Linguistically, the candidates for the domain of *distinction* are the prefixes *non-, a-, dis-, un-* and *in-*. Even though the prefixes have to do with distinction, they are not alike in behaviour. They differ with reference to the degrees of oppositeness. Using Horn's (2002) criteria, the negative prefixes *non-* and (to a lesser degree) *a-* favour a contradictory reading, which arises from non-gradable bases. I argue that *non-* conspicuously describes a choice between two different plans of action, whereas *a-* noticeably describes a choice between two opposing features of things. In this use, *non-* and *a-* are objective in tone. By contrast, the negative prefixes *dis-*, *un-* and *in-* favour a contrary reading, which arises from gradable bases. I argue that *dis-* occupies the lowest level on the scale of contrariness. It is used mainly to evaluate attitudes of people. In this use, it is unfavourable in character. Likewise, I argue that *un-* occupies a medium level on the scale of contrariness. It is used chiefly to evaluate properties of things. In this use, it is disapproving in character. Finally, I argue that *in-* occupies the highest level on the scale of contrariness. It is used mostly to evaluate properties of situations. In this use, it is critical in character.

Let us now analyse some data to see if the prefixes behave in semantically distinct ways within the domain.

(1) a. The *non-appearance* of the band at the concert was disappointing.
b. There are penalties for *non-compliance* with the fire regulations.
c. They were accused of *non-observance* of the terms of the contract.

The sentences cited in (1a-c) contain negative nouns formed by adding the prefix *non-* to nominal bases. In (1a), *non-appearance* means 'not being in a place where people expect to see you'. In (1b), *non-compliance* means 'not obeying a rule'. In (1c), *non-observance* means 'not keeping to a custom'. On scouring the negative nouns, it is found that first the meaning of the prefix is 'dissimilar to the thing imparted by the nominal base', and second that the derived nouns predominantly describe action. This sense is borne out by the collocations of the negative nouns. *Non-appearance* is modified by collocates referring to people such as *adversary, band, ensemble, fan, guide*, etc. *Non-*

compliance is modified by collocates referring to conditions such as *demands, provisions, regulations, requirements, standards*, etc. *Non-observance* is modified by collocates referring to rules such as *agreements, directions, guidelines, stipulations, terms*, etc. More examples of negative nouns include *non-adherence, non-aggression, non-attendance, non-cooperation, non-interference, non-obedience, non-proliferation, non-resistance*, etc.

d. A proportion of the income is spent on *non-essential* goods.
e. They called for the efficient use of *non-renewable* resources.
f. They use *non-standard* dialects for merely literary purposes.

The sentences in (1d-f) contain negative adjectives formed by adding the prefix *non-* to adjectival bases of classifying nature. In (1d), *non-essential* means 'not completely necessary'. In (1e), *non-renewable* means 'cannot be replaced after use'. In (1f), *non-standard* means 'not having been standardised'. On combing the negative adjectives, it is found that first the meaning of the prefix is 'different from the thing imparted by the adjectival base', and second the derived adjectives substantially describe non-technical features of things. This sense is evidenced by the collocations of the negative adjectives. *Non-essential* is accompanied by collocates having to do with production such as *goods, items, products, services, workers*, etc. *Non-renewable* is accompanied by collocates having to do with means such as *energy, fuels, oil, resources, sources*, etc. *Non-standard* is accompanied by collocates having to do with language such as *dialects, forms, phraseology, varieties, versions*, etc. More examples of negative adjectives include *non-alcoholic, non-aligned, non-existent, non-refundable, non-residential, non-returnable, non-stop, non-verbal*, etc.[1]

(2) a. They promote wound healing by adherence to *aseptic* technique.
b. The use of octaves is rare, a common feature of all *atonal* music.
c. *Atypical* patterns emerged from our analysis of the accident data.

The sentences mentioned in (2a-c) contain negative adjectives formed by adding the prefix *a-* to adjectival bases of classifying nature. In (2a), *aseptic* means 'not having harmful bacteria'. In (2b), *atonal* means 'not written in any musical key'. In (2c), *atypical* means 'not typical or usual'. A detailed analysis of the negative adjectives shows that first the meaning of the prefix is 'unlike the quality of thing referred to by the adjectival base', and second the derived adjectives chiefly describe technical features of things. This sense is confirmed by the collocations of the negative adjectives. *Aseptic* co-occurs with collocates referring to medical methodology such as *conditions, modes, procedures, techniques, ways*; medical equipment such as *agent, bandage, dressing, lotion, plaster*, etc. *Atonal* co-occurs with collocates referring to music such as *fuss, guitar, music, noise, variations*, etc. *Atypical* co-occurs with collocates referring to *cases, examples, patterns, results, species*, etc. More examples of negative adjectives include *asymmetric, atemporal*, etc.

(3) a. The guests didn't wish to appear *discourteous*.
    b. Beware of *dishonest* traders in the tourist areas.
    c. He was accused of being *disloyal* to the party.

The sentences given in (3a-c) contain negative adjectives formed by adding the prefix *dis-* to adjectival bases of qualitative nature. In (3a), *discourteous* means 'not courteous, having bad manners'. In (3b), *dishonest* means 'not honest; intending to deceive people'. In (3c), *disloyal* means 'not loyal, not faithful to one's friends, family or country'. Analysis of the negative adjectives has two repercussions. First, the meaning of the prefix is 'the converse of the quality signified by the adjectival base'. Second, the derived adjectives often describe attitudes of people. This sense is validated by the collocations of the negative adjectives. *Discourteous* is associated with collocates denoting people such as *guests, people, relatives, teams, visitors*, etc. *Dishonest* is associated with collocates denoting people such as *dealers, politicians, proprietors, servants, traders*, etc. *Disloyal* is associated with collocates denoting people such as *followers, officers, players, soldiers, users*, etc. Examples of other negative adjectives include *disagreeable, disingenuous, disinterested, disobliging, disquiet*, etc.[2]

(4) a. The industrial premises in the nearby town are *unclean*.
    b. The small front garden is *untidy*; it is full of bushes.
    c. The event was so *unusual* that people crowded the streets.

The sentences in (4a-c) contain negative adjectives formed by adding the prefix *un-* to adjectival bases of qualitative nature. In (4a), *unclean* means 'not clean or sterile'. In (4b), *untidy* means 'not neat or well arranged'. In (4c), *unusual* means 'not usual, different from what is usual or normal'. A review of the negative adjectives has two consequences. First, the meaning of the prefix is 'the antithesis of what is specified by the adjectival base'. Second, the derived adjectives mostly describe properties of things.[3] This sense is verified by the collocations of the negative adjectives. *Unclean* links with collocates naming places such as *city, hotel, office, premise, room*; cloth such as *bandage, dressing, sheet, socks, towel*, etc. *Untidy* links with collocates naming objects such as *beard, bed, desk, hair, moustache*; places such as *bridge, garden, house, nest, room*, etc. *Unusual* links with collocates naming events such as *event, occasion, story, venture;* activity such as *behaviour, hobby, method, request, skill*, etc. Examples of other negative adjectives include *unclear, uncommon, unkempt, unnecessary, unsafe*, etc.[4]

(5) a. Their comments were wholly *inappropriate.*
b. The grammatical patterns are highly *irregular.*
c. The contributions are statistically *insignificant.*

The sentences mentioned in (5a-c) contain negative adjectives formed by adding the prefix *in-* to adjectival bases of qualitative nature. In (5a), *inappropriate* means 'not suitable for a particular situation or occasion'. In (5b), *irregular* means 'not normal; not according to the usual rules'. In (5c), *insignificant* means 'not big or valuable enough to be considered important'. A study of the negative adjectives has two ramifications. First, the meaning of the prefix is 'the opposite of the thing expressed by the adjectival base'. Second, the derived adjectives commonly describe properties of situations.[5] This sense is supported by the collocations of the negative adjectives. *Inappropriate* has collocates describing activities such as *behaviour, comment, discussion, remark, statement,* clothes such as *array, costume, finery, guise, raiment*, etc. *Irregular* collocates with words describing language such as *nouns, patterns, structures, verbs, words*, etc. *Insignificant* occurs with collocates describing measurement such as *feature, level, measure, portion, proportion*; statistics such as *contributions, costs, figures, numbers, payments*, etc. Examples of other negative adjectives include *inaccurate, inadequate, incomplete, illogical, impossible, irrational, irrelevant*, etc.

From the foregoing discussion, we can draw some conclusions. One is that the core meaning of the prefixes is 'not', paraphrased alternatively as dissimilar, different, unlike, the converse of, the antithesis, or the opposite of. Another conclusion is that the examples denote *distinction*, the act of showing the difference between two entities. Evidence in support of the analysis can, as Horn (1989: 281–2) mentions, be formulated in relation to the use of adverbs of degree such as *awfully, comparatively, downright, extremely, fairly, quite, rather, relatively, somewhat, very*, etc. With these adverbs, it seems that *non-* and *a-* derivations are inappropriate, as in **rather non-appearance* or **very aseptic*. The reason is that these adverbs imply gradability, and so are at odds with bases having non-scalar properties, which gives rise to contradictory readings. By contrast, *dis-*, *un-* and *in-* derivations seem to be appropriate, as in *extremely discourteous*, *somewhat unclear* or *quite illegal*. The reason is that these adverbs imply gradability, and so correlate with bases allowing some sort of intermediate state, which gives rise to contrary readings.

Figure 22 presents a configuration which captures the different facets of the domain of *distinction*.

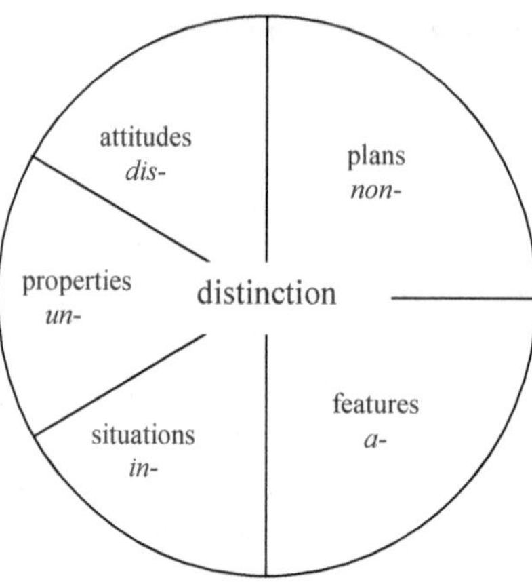

Figure 22: The domain of distinction

4.4.2 The domain of opposition

The domain of *opposition* is a conceptual area which involves the act of opposing something verbally or physically. A verbal opposition relates to the mind, involving the expression of ideas against something in words. A physical opposition relates to the body, involving the use of a means to prevent something from happening, or taking the necessary steps to change it. As the definition discloses, opposition involves three facets. These facets have semantic relationships with one another. The first facet embodies two things. One is attitude, the way you think and feel about a given policy or strategy. Here, the opposition takes the form of a verbal action. Another is substance, the use of a means to prevent something from succeeding. Here, the opposition takes the form of a procedural action. The second facet related to event, the process of doing something in order to deal with a situation. Here, the opposition takes the form of a physical action. The third facet includes contrast, the fact of comparing or putting close together two or more things in order to show the differences between them.

Linguistically, the domain of *opposition* is realised by the negative prefixes *anti-, counter-* and *contra-*. Although these prefixes are similar in symbolising the act of opposing, they are not identical in use. Each stands, I argue, for a different facet of opposition. The negative prefix *anti-* represents the first facet. It refers either to an attitude, where the opposition is against opinions which spark disagreement, or to a device, where the opposition is against states that have harmful effects. The negative prefix *counter-* represents the second facet. It refers to an action or event carried out in retaliation for another action or event. The opposition is against actions or practices that are controversial. It involves attempts to change a state of affairs like acting, fighting, working, etc. The negative prefix *contra-* represents the third facet. It places or sets two things in comparison for the sake of clarity. The opposition is about two things being as different as possible.

Let us now have a check on some data to see if the prefixes have different uses within the domain.

(6) a. *Anti-colonialism* is a powerful theme in modern Asia.
b. The banners were aimed at the *anti-immigration* policy.
c. He played a leading role in the *anti-slavery* movement.

In (6a-c), the negative nouns are formed by adding *anti-* to nominal bases. In (6a), *anti-colonialism* means 'hostile to the belief in and support for the system of one country controlling another'. In (6b), *anti-immigration* means 'hostile to the activity whereby people come to a different country in order to live there permanently'. In (6c), *anti-slavery* means 'hostile to the practice whereby people own other people and force them to work for them'. Careful examination of the examples shows that first the meaning of the prefix is 'reacting against the thing named by the nominal base', and second the derived nouns describe attitudes of people against certain practices or themes, used largely in philosophical, political or social fields. This sense is underlined by the collocations of the negative nouns. *Anti-colonialism* collocates with words such as *argument, concept, policy, stance, theme*, etc. *Anti-immigration* collocates with words such as *bills, laws, ordinances, policies, proposals*, etc. *Anti-slavery* collocates with words such as *lobby, movement, organisation, party, society*, etc. Other negative nouns which share the same meaning include *anti-abortion, anti-discrimination, anti-establishment, anti-government, anti-imperialism*, etc.

d. She is thinking of having the HIV *antibody* test.
e. Heparin is a drug used in *anti-coagulant* treatment.
f. *Anti-histamine* tablets help to relieve the condition.

In (6d-f), the negative nouns are formed by adding *anti-* to nominal bases. In (6d), *an anti-body* is 'a protein produced in the blood to fight antigen'. In (6e), *an anti-coagulant* is 'a substance use to prevent the blood from clotting'. In (6f), *an anti-histamine* is 'a drug used to counteract the effects of histamine, used in treating allergies'. Thorough examination of the examples indicates that first the meaning of the prefix is 'preventing the thing named by the nominal base', and second the derived nouns describe devices used against illnesses, mainly in the medical field. This sense is highlighted by the collocations of the negative nouns. *Antibody* collocates with words such as *evidence, injection, molecule, production, test*, etc. *Anti-coagulant* collocates with words such as *agent, drug, therapy, treatment*, etc. *Anti-histamine* collocates with words such as *cream, drops, injections, shots, tablets*, etc. Other negative nouns with the same meaning include *anti-cancer, anti-infection, anti-cholesterol, anti-depressant, anti-virus*, etc.

g. They were asked to hand in their *anti-aircraft* guns.
h. The aircraft is fitted with a new *anti-missile* system.
i. They plan for a new generation of *anti-tank* weapons.

In (6g-i), the negative nouns are formed by adding *anti-* to nominal bases. In (6g), an *anti-aircraft gun* is 'a weapon designed to destroy enemy aircraft'. In (6h), an *anti-missile system* is 'a weapon designed to destroy enemy missiles'. In (6i), *anti-tank* weapons are 'weapons designed to destroy enemy tanks'. Close scrutiny of the examples demonstrates that first the meaning of the prefix is 'defending against the weapon named by the nominal base', and second the derived nouns describe weapons used against the enemy, mostly in the military field. This sense is accentuated by the collocations of the negated nouns. *Anti-aircraft* collocates with words such as *artilleries, batteries, frigates, guns, missiles*, etc. *Anti-missile* collocates with words such as *lasers, launchers, rockets, satellites, systems*, etc. *Anti-tank* collocates with words such as *bombs, guns, mortars, tubes, weapons*, etc. Other negative nouns which undergo the same analysis include *anti-ballistic, anti-personnel, anti-ship, anti-submarine*, etc.

(7) a. The armed forces launched a rapid *counter-attack* against the rebels.
b. They adopted specific *counter-measures* against a threatened strike.
c. They gathered the necessary forces for a military *counter-offensive*.

In (7a-c), the negative nouns are formed by adding *counter-* to nominal bases. In (7a), *a counter-attack* means 'an attack made in response to the attack of an enemy or opponent in war, sport or an argument'. In (7b), *counter-measure* means 'an action taken to deal with a threat or as a defence against a hostile action by somebody else'. In (7c), *a counter-offensive* means 'an attack made in order to defend against enemy attacks'. Careful scrutiny of the instances reveals that first the prefix conveys the meaning of 'facing the action or activity designated by the nominal base', and second the derived nouns principally describe actions taken against opposite actions, used mainly in military or sporting fields. This sense is emphasised by the collocations of the negative nouns. *Counter-attack* collocates with words such as *effective, instant, rapid, rigorous, swift*, etc. *Counter-measure* collocates with words such as *effective, immediate, solid, specific, thermal*, etc. *Counter-offensive* collocates with words such as *air, land, military, naval, sea*, etc. Other negative nouns which involve the same analysis include *counter-act, counter-march, counter-espionage, counter-punch, counter-strike*, etc.

(8) a. Constitutionalism is set in direct *contradistinction* to arbitrary power.
b. The strong contraindication convinced the patient not to take the pills.
c. They took up a stand in logical *contraposition* to government policies.

In (8a-c), the negative nouns are formed by adding *contra-* to nominal bases. In (8a), *contradistinction* means 'in contrast with something or somebody'. In (8b), *contraindication* means 'a possible reason for not giving somebody a particular drug or medical treatment'. In (8c), *contraposition* means 'a position opposite to or against something'. An investigation of the instances explains that first the prefix incorporates the sense of 'posed against the thing mentioned in the nominal base', and second the derived nouns basically describe the contrast between two opposing things, used in general contexts. This is indicated by the collocations of the negative nouns. *Contradistinction* collocates with words such as *conscious, direct, forthright, plain, sharp*, etc. *Contraindication* collocates with words such as *absolute, complete, relative, specific, strong*, etc. *Contraposition* collocates with words such as *literal, logical, non-literal, rational, suitable*, etc.

From the preceding discussion, we can reach some conclusions. One conclusion is that the core meaning of the prefixes is 'in response to', paraphrased alternatively as reacting, preventing, defending, facing, or comparing. Another conclusion is that the examples denote *opposition*, the act of disagreeing with somebody or something. Evidence in support of the *anti-* derivatives comes from the fact that they can be used with expressions designating attitude. It is felicitous to say *an anti-immigration policy*, but it is infelicitous to say **an anti-immigration offensive*. This is because a noun negated by the prefix *anti-* collocates only with expressions referring to attitude as shown by *policy*, and not action as shown by *offensive*. Likewise, it is possible to say *an anti-aircraft gun*, but it is impossible to say **an anti-aircraft attack*. The reason is that the prefix *anti-* tends to collocate with words referring to equipment like *gun*, and not action like *attack*. By contrast, evidence in support of the *counter-* derivatives comes from the fact that they can be used with expressions modifying action. It is right to say *an instant counter-attack*, but it is wrong to say **an instant anti-immigration policy*. The reason is that of the two prefixes only *counter-* accepts the company of words modifying action like *instant*.

Figure 23 presents a configuration which captures the different facets of the domain of *opposition*.

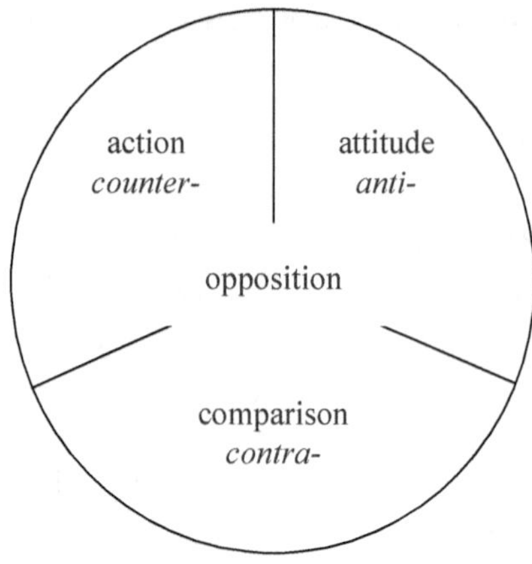

Figure 23: The domain of opposition

4.4.3 The domain of privation

The domain of *privation* is a conceptual area involving the relationship between two entities in which one suffers from the absence of something, usually not positive. It refers to the fact of not having the basic things that one considers necessary. The focus is on the act of taking something away from something, preventing somebody from having something, or the state where something is lacking or has been denied. In logic, it indicates the absence of any quality which might be naturally or rationally expected. Privation has three facets. The first facet signifies action. It entails harming, confiscating or reducing rights of people, especially as a punishment. The second facet signifies something concrete. It means damaging, messing up or neglecting a place. The third facet signifies something abstract. It involves something important which is lacking in a particular situation.

Linguistically, prefixes in English that are a manifestation of the domain of *privation* are *de-*, *dis-*, and *un-*. These prefixes have one thing in common. They share reference to the act of disowning. However, they are not interchangeable. Each prefix represents, I argue, a certain semantic specialisation of the domain. They differ with respect to the type of entity being affected in the process. The negative prefix *de-* represents the semantic value of the first facet. It reflects on things or places which lack certain qualities. The negative prefix *dis-* signifies the semantic value of the second facet. It focuses on people who are deprived of certain qualities. The negative prefix *un-* symbolises the semantic value of the third facet. It places emphasis on situations which suffer from the absence of certain properties.

Let us now run a check on some data to see if the prefixes are associated with distinct patterns within the domain.

(9) a. The new proposal is meant to *debase* the currency.
 b. The disease had *deformed* the spine of the patient.
 c. They were charged with *defacing* public property.

The negative verbs in (9a-c) are formed by appending *de-* to nominal bases. In (9a), *debase* means 'to reduce the quality or value of something'. In (9b), *deform* means 'to change or spoil the usual or natural shape of something'. In (9c), *deface* means 'to damage the appearance of something especially by drawing or writing on it'. A quick trawl through the examples yields two facts. First, the meaning of the prefix is 'depriving the thing described by the nominal base'. Second, the derived verbs take things to be suffering from the deprivation of certain qualities.[6] This sense is asserted by the collocations of the negative nouns. *Debase* collocates with words such as *coinage, currency, money, surroundings, work*, etc. *Deform* collocates with words such as *body, foot, hand, joint, spine*, etc. *Deface* collocates with words such as *garden, home, item, property, wall*, etc. Other negative verbs which accommodate to this meaning include *decry, delimit, declaim, demark, despoil*, etc.

(10) a. Her behaviour has inflicted *dishonour* upon the family.
 b. She seems to have fallen into *disfavour* with the director.
 c. The minister's comments do teachers a great *disservice*.

The negative nouns in (10a-c) are formed by appending *dis-* to nominal bases. In (10a), *dishonour* means 'the loss of other people's respect and approval because of the bad way somebody has behaved'. In (10b), *disfavour* means 'the feeling that you do not like or approve of somebody or something'. In (10c), *disservice* means 'to do something that harms somebody and the opinion that other people have of them'. A brief perusal of the examples ends in two results. First, the meaning of the prefix is 'lacking the thing signified by the nominal base'. Second, the derived nouns show people to be suffering from certain qualities.[7] This sense is backed up by the collocations of the negative nouns. *Disgrace* collocates with words such as *clan, family, leader, parents, politician*, etc. *Disfavour* collocates with words such as *director, fellows, headmaster, investors, workers*, etc. *Disservice* collocates with words such as *employees, farmers, litigants, scholars, teachers*, etc. Other negative nouns which share this meaning include *discomfort, disgrace, disharmony, disinterest, disrepute, disrespect*, etc.

(11) a. Illness is a feeling of mental *unhealth* which is entirely personal.
 b. Because of injustice, the world is in a state of regional *unpeace*.
 c. He stressed the culpable *unwisdom* of fathers leaving their kids.

The negative nouns in (11a-c) are formed by appending *dis-* to nominal bases. These nouns are very infrequent, and many speakers would not accept them in fact. In (11a), *unhealth* means to 'a lack of fitness, soundness or wellness'. In (11b), *unpeace* means 'a lack of accord, peace or stability'. In (11c), *unwisdom* means 'a lack of wisdom, judgement or good sense'. A close look at the examples produces two effects. First, the meaning of the prefix is 'bereft of what is specified by the nominal base'. Second, the derived nouns construe situations as suffering from certain qualities. This sense is upheld by the collocations of the negative nouns. *Unhealth* collocates with words such as *bodily, emotional, mental, physical, spiritual*, etc. *Unpeace* collocates with words such as *financial, international, marital, political, regional*, etc. *Unwisdom* collocates with words such as *conventional, culpable, political, rash, utter*, etc. Other negative nouns which share this meaning include *unconcern, unease, unlaw, unrest, untruth*, etc.

The analysis conducted so far leads to the following conclusions. Firstly, the core meaning of the prefixes is 'strip', paraphrased alternatively as depriving, lacking or bereft of. Another conclusion is that the examples denote *privation*, the act of causing somebody or something to lack something. Evidence in support of the analysis may be gained from the collocations that accompany such words. It is correct to say *debase something*, but it is incorrect to say **debase a situation*. This is because in expressing privation *de-* acts on things for the sake of deprivation. Likewise, it is appropriate to say *disgrace someone*, but it is inappropriate to say **disgrace something*. This is because in expressing privation *dis-* picks out people for the sake of deprivation. Finally, it is accurate to say *a period of unrest,* but it is inaccurate to say * *a period of debase* or * *a period of disgrace*. The reason is that in expressing privation *un-* construes abstract situations as being in the process of deprivation.

Figure 24 presents a configuration which captures the different facets of the domain of *privation*.

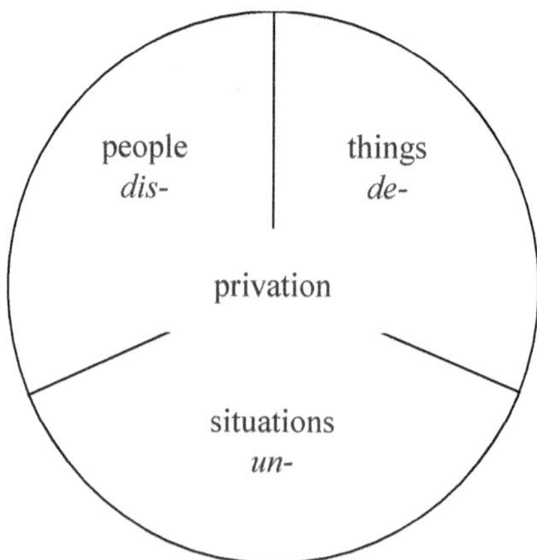

Figure 24: The domain of privation

4.4.4 The domain of removal

The domain of *removal* is a conceptual area involving taking away of something, getting rid of something, or changing the position of someone. It is the act of removing or the fact of being removed. As the definition reveals, removal subsumes different processes in accordance with the different entities being removed. One process involves taking something or someone away from somewhere, or off something. Another process involves forcing someone to leave an important job or a position of power because they have behaved badly or not in a way you approve of. A further process involves freeing, relieving, or emptying a place or thing of something, usually something undesirable. Relative to the nature of the entity affected in the process, removal comprises three facets. Firstly, things or places are being affected in the act of removing. Secondly, people are being affected in the act of removing. And thirdly, physical objects are being affected in the act of removing.

Linguistically, the domain of *removal* is coded by the negative prefixes *de-*, *dis-* and *un-*. In general terms, these prefixes are alike in expressing the notion of removal. They convey the core meaning of 'remove'. They express a shift in position, change in content or do away with something. In specific terms, however, they are different. Each prefix symbolises, I argue, a different facet of the domain. Each has, therefore, a particular use. The distinction lies in the type of object which undergoes the process of removal. The prefix *de-* stands for the first facet. It chooses places or things that are part of objects in the act of removal. The prefix *dis-* represents the second facet. It opts for people in the act of removal. The prefix *un-* represents the third facet. It selects physical objects in the act of removal.

Let us now peruse some data to see if the prefixes really have different usages within the domain.

(12) a. They *degreased* the engine.
b. They *defrosted* the freezer.
c. The *decontrolled* the market.

The sentences in (12a-c) include negative verbs formed by attaching *de-* to verbal bases. (12a), *degrease* means 'to remove grease from something'. In (12b), *defrost* means 'to remove frost from something'. In (12c), *decontrol* means 'to remove control from something'. A thorough examination of the examples tells us two things. First, the meaning of the prefix is 'removing the thing described by the nominal base'. Second, the derived verbs describe things regardless of whether they are abstract or concrete.[8] This sense is supported by the collocations of the negative verbs. *Degrease* collocates with words such as *cover, engine, machinery, skin, surface*, etc. *Defrost* collocates with words such as *chicken, freezer, fridge, sandwich, store*, etc. *Decontrol* collocates with words such as *currency, dwellings, economy, market, price*, etc. Additional examples of negative verbs include *debug, defog, defuse, degas, dehorn, de-ice, demist*, etc.

(13) a. Most of the rebels were captured and *disarmed.*
b. The actress went behind the screen to *disrobe.*
c. Building the dam will *displace* all the villagers.

The sentences in (13a-c) include negative verbs formed by attaching *dis-* to nominal bases. In (13a), disarm means 'to take weapons away from someone'. In (13b), disrobe means 'to take off your or somebody else's clothes'. In (13c), *displace* means to 'remove somebody from an original position'. A full investigation of the examples tells us two things. First, the meaning of the prefix is 'ridding of the thing signified by the nominal base'. Second, the derived verbs describe people.[9] This sense is upheld by the collocations of the negative verbs. *Disarm* collocates with words such as *contras, guerrillas, rebels, fighters, troops*, etc. *Disrobe* collocates with words describing people. *Displace* collocates with words such as *citizens, people, population, settlers, villagers*, etc. Additional examples of negative verbs include *disconcert, disembowel, disenfranchise, dishearten, dispossess*, etc.

(14) a. He watched the cruel master *uncurling* the whip.
b. The imperial pigeon can *unhitch* its lower beak.
c. They docked in the river to *unload* their cargoes.

The sentences in (14a-c) include negative verbs formed by attaching *un-* to nominal bases. In (14a), *uncurl* means 'to straighten from a curled position'. In (14b), *unhitch* means 'to detach something held by a hook'. In (14c), *unload* means 'to remove the contents of something, especially a load of goods from a vehicle'. A full exploration of the examples shows two things. First, the meaning of the prefix is 'taking away what is specified by the nominal base'. Second, the derived verbs describe physical objects.[10] This sense is supported by the collocations of the negative verbs. *Uncurl* collocates with words such as *bone, finger, flag, lip, whip*, etc. *Unhitch* collocates with words such as *beak, caravan, trailer, van, wagon*, etc. *Unload* collocates with words such as *boxes, cargoes, crates, provisions, sacks*, etc. Additional examples of negative verbs include *uncage, unchain, unharness, unhook, unmask*, etc.

Let us now present the main conclusions of the argument. One conclusion is that the core meaning of the prefixes is 'remove', paraphrased alternatively as releasing, ridding or taking away. Another conclusion is that the examples denote removal, the act of getting rid of somebody or something. A piece of evidence supporting the analysis of a base negated by *de-* is that it cannot be modified by words referring to people as in **degrease a rebel*, or objects as in **degrease a whip*. A piece of evidence supporting the analysis of a base negated by *dis-* is that it cannot be modified by words referring to things or places as in **disarm a market*, or objects as in **disarm a whip*. A piece of evidence supporting the analysis of a base negated by *un-* is that it cannot be modified by words referring to places as in **uncurl a market*, or people as in **uncurl a rebel*.

Figure 25 presents a configuration which captures the different facets of the domain of *removal*.

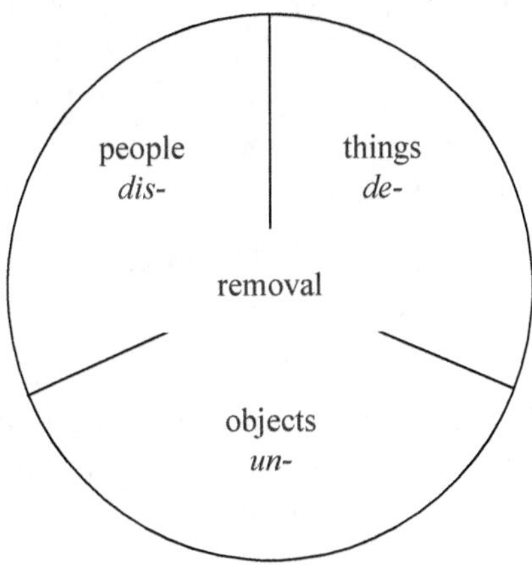

Figure 25: The domain of removal

4.4.5 The domain of reversal

The domain of *reversal* is a conceptual area involving the relationship between two entities which witness a change from one state to its opposite. It encodes a change of something so that it is the opposite of what it was before. It is the act of turning in the opposite direction or position, or changing the usual order of the parts of something. It manifests itself in different ways relative to the particular area of meaning in which it is operating. Depending on the nature of the object being reversed, reversal comprises three facets. In the first, the act of reversing relates to places. In this way, it refers to an attempt to make a place different from how it was. In the second, the act of reversing pertains to humans. In this way, it refers to an attempt to make somebody suffer from the opposite effect of an action. In the third, the act of reversing applies to physical objects. In this way, it refers to an attempt to make something return to the opposite position from that which it has previously been.

Linguistically, the domain of *reversal* is coded by the negative prefixes *de-*, *dis-* and *un-*, each of which is required to describe a different facet. To the extent that these prefixes pertain to the domain of reversal, they are similar.

The prefixes have the meaning 'undo' as their main semantic component. The use of a reversal prefix is to indicate that the affected entity returns to the state obtaining before the process expressed in the base took place. Even so, each prefix is associated, I argue, with a particular facet of the domain, and has therefore a distinctive meaning. The distinction resides in the type of object which the prefix selects to affect in the process of reversing. The prefix *de-* stands for the first facet: it targets places or things in the act of reversal. The prefix *dis-* stands for the second facet: it subjects people to the act of reversal. The prefix *un-* stands for the third facet: it exposes physical objects to the act of reversal.

Let us now carry out a check on some data to see if the prefixes occur in different environments within the domain.

(15) a. Can you *decipher* the writing on this envelope?
b. The soldiers deliberately *defiled* the holy places.
c. The corporation *deforested* large areas of forest.

Exemplified in (15a-c) are sentences which contain negative verbs formed by attaching *de-* to verbal bases. (15a), *decipher* means 'to discover the meaning of something written badly or in a difficult or hidden way'. In (15b), *defile* means 'to make a place, especially a holy place, dirty or no longer pure'. In (15c), *deforest* means 'to cut down and destroy all the trees in a place'. A careful review of the examples identifies two outcomes. First, the meaning of the prefix is 'reversing the action described by the nominal base'. Second, the derived verbs describe places or things. This sense is supported by the collocations of the negative verbs. *Decipher* collocates with words such as *jottings*, *messages*, *signs*, *wordings*, *writing*, etc. *Defile* collocates with words such as *areas, buildings, places, surroundings, walls*, etc. *Deforest* collocates with words such as *area, forests, land, timberland, woods*, etc. Examples of other negative verbs include *decentralise, decode, demilitarise, depolarise, desegregate*, etc.

(16) a. The contras have consistently rejected calls to *disband.*
b. The photos were deliberately taken to *discredit* the boss.
c. He *disinherited* his daughter because she married Stephen.

Exemplified in (16a-c) are sentences which contain negative verbs formed by attaching *dis-* to verbal bases. In (16a), *disband* means 'to stop people from operating as a group'. In (16b), *discredit* means is 'to make people stop respecting a person or his/her idea'. In (16c), *disinherit* means 'to prevent somebody, especially one's son or daughter, from receiving one's money or property after death'. A detailed study of the examples confirms two points. First, the meaning of the prefix is 'turning around the action signified by the verbal base'. Second, the derived verbs describe people. This sense is corroborated by the collocations of the negative verbs. *Disband* collocates with words such as *army, contras, council, police, rebels*, etc. *Discredit* collocates with words such as *boss, enemy, politician, reformer, witness*, etc. *Disinherit* collocates with words such as *children, daughter, family, heir, son*, etc. Examples of other negative verbs include *displace, disempower, dishearten, displease, dissociate*, etc.

(17) a. She was trembling as she tried to *unlock* the door.
b. He managed to *unscrew* the hinges of the windows.
c. She wanted to *unwrap* the package in his presence.

Exemplified in (17a-c) are sentences which contain negative verbs formed by attaching *un-* to verbal bases. In (17a), *unlock* means 'to undo the lock of something using a key'. In (17b), *unscrew* means 'to undo the screw of something by turning it round'. In (17c), *unwrap* means 'to undo the wrap of something by opening it'. In searching the examples, we can make two comments. First, the meaning of the prefix is 'inverting what is specified by the verbal base'. Second, the derived verbs describe objects. This sense is substantiated by the collocations of the negative verbs. *Unlock* collocates with words such as *cupboard, door, gate, safe, window*, etc. *Unscrew* collocates with words such as *cap, hinge, lid, nut, tap*, etc. *Unwrap* collocates with words such as *coil, gift, package, roll, strip*, etc. Examples of other negative verbs include *unbend, uncurl, unfold, unplug, unroll*, etc.[11]

Before ending the discussion, we turn to the main conclusions. One conclusion is that the core meaning of the prefixes is 'undo', paraphrased alternatively as reversing, turning around or inverting. Another conclusion is that the examples denote reversal, the act of turning something the opposite way around. Evidence in support of the analysis relates to the (in)compatibility of the negatively-prefixed forms with the categories of collocations accompanying them. For example, it is right to say *They defiled the place*, but it is faulty to say **They defiled the rebels*. The reason is that only *de-* is compatible with words referring to places. Likewise, it is appropriate to say *They disbanded the rebels*, but it is inappropriate to say **They disbanded the place*. The reason is that only *dis-* is suitable with words denoting people. Finally, it is permissible to say *He unlocked the door*, but it is not permissible to say **He unlocked the rebels*. The reason is that only *un-* is only used to describe objects.

Figure 26 presents a configuration which captures the different facets of the domain of *reversal*.

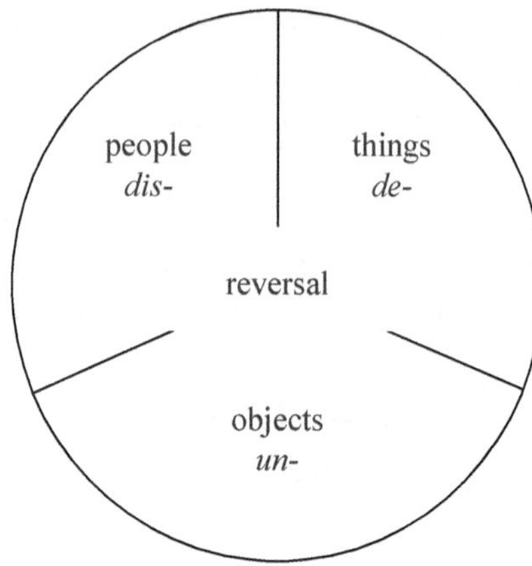

Figure 26: The domain of reversal

4.4.6 The domain of treatment

The domain of *treatment* is a conceptual area which encodes a way of behaving towards or dealing with a person or thing. As an important human experience, *treatment* covers two facets of behaviour, depending on whether the action taken is a matter of accidental or intentional. One facet revolves around accidental treatment, where an action happens by chance. It is an incorrect, unwise or unfortunate act caused by poor judgement or lack of care. It is an action that is liable to happen as a consequence of something else. The thing affected is either a thing or an object. Accordingly, the performer of the action is not subjected to punishment if s/he does it only once. The other facet centres on intentional treatment, where an action is carefully planned or consciously done. It is an act of reflecting, planning or arranging something in advance. The thing affected involves a process that is done on purpose. Accordingly, the performer is held responsible for his or her action.

Linguistically, the domain of *treatment* is actualised by the negative prefixes *mis-* and *mal-*. Both prefixes are associated with the sense of 'wrong', but their applications may be distinguished. Each prefix represents, I argue, a distinct aspect of the domain. Each has an individual role to play in the language. Their distinctiveness becomes apparent when they are used in context. The prefix *mis-* represents the accidental facet. It means handling somebody or something wrongly out of negligence. It stresses an action that is incorrectly or unconsciously done. As evidence, one can say *mishear* but not *mislisten* because *listen* is a verb of endeavour and denotes conscious performance. By contrast, the prefix *mal-* represents the intentional facet. It means handling a person or an animal intentionally cruelly or unkindly or handling a thing out of personal interest. It stresses an action that is purposely or knowledgeably performed.

Let us now check the data to see if the prefixes have different semantic preferences within the domain.

(18) a. Harris had *misconstrued* her words.
b. They *misinterpreted* the regulations.
c. The newspaper *misreported* the facts.

In the examples under (18a-c), *mis-* is used with verbal bases to form negative verbs. In (18a), *misconstrue* means 'to understand somebody's words or actions wrongly'. In (18b), *misinterpret* means 'to understand somebody's words wrongly'. In (18c), *misreport* means 'to give a report of an event, etc. that is not correct'. Browsing through the examples leads us to two conclusions. First, the prefix imparts the meaning of 'falsely performing the action conveyed by the verbal base'. Second, the derived verbs overwhelmingly describe mental acts. This sense is attested to by the collocations of the negative verbs. *Misconstrue* collocates with words such as *act, comment, remarks, story, word*, etc. *Misinterpret* collocates with words such as *account, expression, question, regulation, statement*, etc. *Misreport* collocates with words such as *cases, effects, facts, findings, incidents*, etc. Other negative verbs which reflect this meaning include *misapprehend, miscalculate, misconceive, misconstruct, misinform, misjudge, misquote, misread, misrepresent, misrule*, etc. *Mis-* can be equally used with nouns derived from these verbs, as in *I was under the misapprehension that the course was for complete beginners*, *A popular misconception is that plastic is a hazard to the environment*, and *Any misconstruction of these terms could lead to errors of law*.

d. He *mishit* the ball which flew into an adjacent court.
e. Could I borrow a pen? I seem to have *mislaid* mine.
f. You can obtain a replacement if you *misplace* the envelope.

In the examples under (18d-f), *mis-* is used with verbal bases to form negative verbs. In (18d), *mishit* means 'to hit the ball wrongly so that it does not go where one had intended'. In (18e), *mislay* means 'to lose something temporarily by forgetting where you have put it' In (18f), *misplace* means 'to put something somewhere and then be unable to find it again, especially for a short time'. Scanning through the examples leads us to two conclusions. First, the prefix imparts the meaning of 'falsely performing the action conveyed by the verbal base'. Second, the derived verbs primarily describe physical objects. This sense is testified to by the collocations of the negative verbs. *Mishit* collocates with words such as *ball, corner, kick, pass, shot*, etc. *Mislay* collocates with words such as *car, key, mobile, pen, things*, etc. *Misplace* collocates with words such as *envelope, key, letter, towel, wallet*, etc. Other negative verbs which manifest this meaning include *misapply, misdirect, mischarge, miskick, misthrow*, etc. *Mis-* can equally be used with nouns derived from these verbs, as in *He was accused of the misappropriation of the company funds*, and *The project collapsed through financial mismanagement*.

g. The pupil was expelled for persistent *misbehaviour*.
h. The doctor was accused of professional *misconduct*.
i. The children were rebuked for minor *misdemeanours*.

In the examples under (18g-i), *mis-* is used with nominal bases to form negative nouns. In (18g), *misbehaviour* means 'unacceptable or inappropriate behaviour, especially by children'. In (18h), *misconduct* means 'unacceptable behaviour, especially by professional people'. In (18i), *misdemeanour* means 'unacceptable, but not very serious, action'. In skimming through the examples, we can make two comments. First, the prefix imparts the meaning of 'false performance of the thing conveyed by the nominal base'. Second, the derived nouns frequently describe behaviour. This sense is certified by the collocations of the negative verbs. *Misbehaviour* collocates with words such as *persistent, raucous, serious, tenacious, wilful*, etc. *Misconduct* collocates with adjectives such as *criminal, financial, gross, judicial, professional*, etc. *Misdemeanour* collocates with words such as *minor, petty, slight, unfortunate, youthful*, etc.[12]

(19) a. They found the authority guilty of repeated *maladministration*.
b. Many of the refugees are suffering from severe *malnutrition*.
c. The clerk is currently standing trial for financial *malpractices*.

In the examples under (19a-c), *mal-* is used with nominal bases to form negative nouns. In (19a), *maladministration* means 'the act of managing something inefficiently or dishonestly'. In (19b), *malnutrition* means 'the act of feeding somebody poorly or inadequately'. In (19c), *malpractice* means 'the act of behaving improperly or unethically, especially by a person in a professional position'. Trawling through the examples results in two observations. First, the prefix imparts the meaning of 'inappropriate execution of the thing denoted by the nominal base'. Second the derived nouns broadly describe action. This sense is maintained by the collocations of the negative nouns. *Maladministration* collocates with words such as *alleged, considerable, occasional, repeated, substantial*, etc. *Malnutrition* collocates with words such as *actual, chronic, severe, slight, widespread*, etc. *Malpractice* collocates with words such as *administrative, electoral, financial, market, medical*, etc. Other negative nouns which share this pattern include *maladjustment, malcontent, malformation, malfunction, maltreatment,* etc.

d. The party is unhappy about the *maladroit* handling of the case.
e. The smell wafting from the place was distinctly *malodorous.*
f. The doctor diagnosed the child as being severely *malnourished.*

In the examples under (19d-f), *mal-* is used with adjectival bases to form negative adjectives. In (19d), *maladroit* means 'the state of being awkward or tactless'. In (19e), *malodorous* means 'the state of being unpleasant or offensive in smell'. In (19f), *malnourished* means 'the state of not being provided with adequate nourishment'. A consideration of the examples leads us to two conclusions. First, the prefix imparts the meaning of 'unsuitably characterised by the thing denoted by the adjectival base'. Second, the derived adjectives generally describe state. This sense is sustained by the collocations of the negative adjectives. *Maladroit* collocates with words such as *handling, interview, relations, version, way*, etc. *Malodorous* collocates with words such as *drains, herds, jacket, places, vegetables*, etc. *Malnourished* collocates with words such as *baby, child, man, patient, population*, etc. Other negative adjectives which fit this pattern include *maladjusted, malformed,* etc.

From the discussion so far, some conclusions emerge. One conclusion is that the core meaning of the prefixes is 'wrong', paraphrased alternatively as false, improper, inappropriate or unsuitable. Another conclusion is that the examples denote treatment, the act of dealing with somebody or something. A piece of evidence supporting the *mis-* analysis is that the sentence can only tolerate non-intentional adverbs such as *accidentally, inadvertently, unexpectedly*, etc. It is acceptable to say *He accidentally misinterpreted our comments*, but it would be odd to say **He intentionally misinterpreted our comments.* This is so because the subject has no power to avoid the occurrence of the action denoted by the base. A piece of evidence supporting the *mal-* analysis is that the sentence can only tolerate intentional adverbs such as *intentionally, consciously, purposely*, etc. It is acceptable to say *The Commission continues to intentionally mal-administer its mandate*, but it is odd to say **The Commission continues to accidentally mal-administer its mandate*. This is so because the subject performs the action denoted by the base knowingly and consciously. In *mis-* derivations, the action is beyond the control of the experiencer, whereas in *mal-* derivations it is under his or her control. The use of such adverbs serve as a confirmation of the semantic values of the words. They have an intensifying effect.

Figure 27 presents a configuration which captures the different facets of the domain of *treatment*.

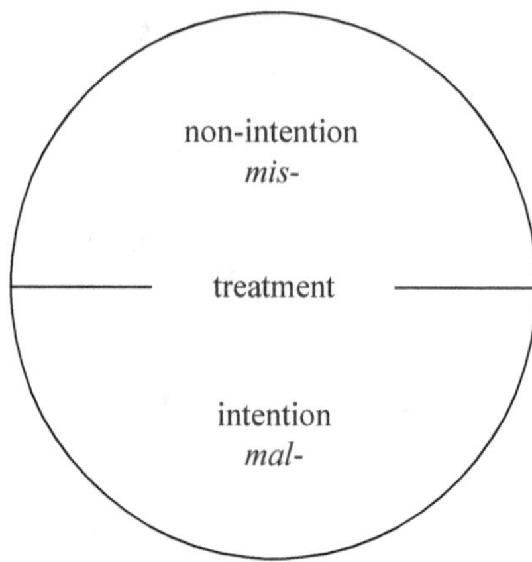

Figure 27: The domain of treatment

4.5 Summary

In this chapter, I have demonstrated three theories according to which linguists describe relationships among meanings of lexemes. Semantic Field theorists believe that the meaning of a lexeme is characterised in terms of the relationship it holds with the other lexemes in the language. Componential Analysis theorists believe that the meaning of a lexeme is characterised by the inherent features it possesses. Cognitive Domain theorists believe that the meaning of a lexeme is characterised in terms of the knowledge structure which it activates. Unlike the Semantic Field theory which describes meaning in terms of paradigmatic and syntagmatic relationships among lexemes, the Cognitive Domain theory describes meaning in terms of a structured background of experience. Lexemes are thus not related to each other directly, but only by virtue of their links to cognitive domains, which provide a conceptual foundation for their meaning. The meaning of a lexeme cannot be understood independently of the vast system of encyclopaedic knowledge to which it is linked. Semantic knowledge is grounded in human interaction with others (social experiences) and with the world around us (physical experiences). Knowledge is represented in the mind as perceptual symbols or mental representations.

A consideration of the issues discussed so far leads to some significant conclusions:

1) Negative prefixes should not be treated in isolation. For their precise description, they should be placed in appropriate cognitive domains, conceptual areas linked to arguments and based on experiences.

Negative prefixes fall into various domains. In addition to *inadequacy* and *degradation*, these are *distinction, opposition, privation, removal, reversal* and *treatment*. Each is an area of activity within which humans are engaged.

2) Each domain has its own internal structure. It comprises particular facets, which deal with diverse features of an experience, and so have discernible characteristics. Negative prefixes keep company by representing the nuclear meaning of a domain, but part company by representing its different facets. The differences in their representational roles are conducive to differences in meaning.

3) The membership of a negative prefix in a domain is determined by its genus, which is based on its definitional analysis. Negative prefixes having more or less the same genus belong to the same domain. Due to polysemy, negative prefixes take part in more than one domain. The prefixes *de-, dis-* and *un-* symbolise *privation, removal* and *reversal.* In each case, the prefix has a separate nuance.

4) The precise meaning of a negative prefix results from identifying the domain in which it is embedded, and from contrasting it with competing counterparts. The task of a cognitive linguist then is to identify the conceptual areas which the prefixes cover and the manner in which their meanings fit the discrete facets. Negative prefixes set therefore a good example of lexical relationships.

5) The characterisations of negative prefixes are conducted in terms of their general patterns of use. Of course, the prefixes overlap in some cases, but the aim is to identify the precise cases of how a negative prefix is often used and what it usually conveys. The exceptions are worth noting, but they do not hinder one from describing, through illustrations and evidence, general lexical behaviour.

Notes

1 A further piece of evidence supporting the descriptive use of the prefix *non-* comes when it is contrasted with the prefix *sub-* which means 'below'. *A substandard accommodation* is one that is below a satisfactory standard. *A non-standard plug* is one that is not normal or usual.

2 In rare cases, the prefix *dis-* is used to describe things and people, as in *The plane disappeared behind a cloud, We saw the booster rockets disengage and fall into the sea*, and *Stress is very disadvantageous in learning situations.*

3 In rare cases, the prefix *un-* is used to describe people, as in *She was unafraid of conflict*, and *He was completely unaware of the whole affair.*

4 In some cases, the derived word has two interpretations. In the expression, *unharnessing a horse*, which is cited in Huddleston & Pullum (2002: 1690), the verb *unharnessing* can be thought of as removing the harness from it or as reversing the process of harnessing it. Another example is the interaction between the domains of *distinction* and *reversal.* In the expression *an unlocked door*, the adjective *unlocked* can be interpreted either *un-locked* where it denotes *distinction*, or *unlock-ed* where it denotes *reversal.*

5 In rare cases, the prefix *in-* is used to describe people, as in *He has described the government as incompetent,* and *He is immodest enough to think that he played an important part in her decision.*

6 In rare cases, the privative *de-* can be attached to bases to describe humans, as in *They denied any intention to defame the senator, He debased his enemies with slanderous insults*, and *They threatened to deskill all the male workers.*

7 In rare cases, the privative *dis-* can be attached to bases to describe places, as in *After the party, the room was in a state of disorder*, and *The station quickly fell into disrepair after it was closed.*

8 In expressing *removal*, *de-* can attach to bases to describe people but rather rarely, as in *The world champion was dethroned by a young challenger*, and *She was deposed as leader of the party in 1991.*

9 In certain formations the prefix *dis-* applies to objects in expressing *reversal*, as in *The earthquake dislodged stones from the walls, The Ottoman Empire disintegrated into lots of small states*, and *I tried to disentangle the wires under my desk.* Similar negative verbs are *disconnect, disfigure, disinfect, disinvest, disuse*, etc.

10 In expressing *removal*, *un-* can attach to bases to describe people but rather rarely, as in *The opposition candidate failed to unseat the cabinet minister*, and *Unhand me, you wretched coward*! and *Constant conflict finally unmanned him.*

11 Horn (2002) lists a number of nonce nouns in *un-*, a representative sample of which includes *unbank, uncola, undate, unfriend, ungalley, unhotel, unplace, unseason, unturkey, unvote, unword*, etc.

12 In certain pairs, the prefix *ill-* competes with the prefix *mis-*. In *The whole ill-conceived scheme was finally abandoned*, the negative adjective *ill-conceived* means badly planned or designed. By contrast, in *It was a misconceived education policy*, the negative adjective *misconceived* means not carefully thought about. In *The public is ill-informed about their legal rights*, the negative adjective *ill-informed* means having or showing little knowledge of something. By contrast, in *It was a misinformed belief*, the negative adjective *misinformed* means that the belief was based on false information.

5 Construal

5.0 Overview

This chapter investigates the third axis of my semantic system, the role of construal in the interpretation of utterances which include negative prefixes. To do this, I validate two tenets of Cognitive Semantics. One tenet is that the morphological structure of an expression reflects its conceptual structure which is derived from experience. Applying this tenet to morphology, I argue that the form of an utterance is determined by its meaning. Another tenet is that morphologically distinct structures reflect differences in meaning of some kind. Applying this tenet to morphology, I argue that the uses of two related utterances serve different needs of communication. The difference between them is a function of construal which the speaker imposes on their common content. To that end, the chapter is organised as follows. Section 1 touches upon the three important elements of communication: message, meaning and interpretation. Section 2 weighs two theories of interpretation. One is the 'meaning equals reference' theory, which treats meaning as existing outside the mind. The other is the 'meaning equals conceptualisation' theory, which treats meaning as existing in the mind. Section 3 applies the conceptualist theory to the interpretation of negatively- prefixed constructions. Section 4 provides the semantic contrasts that set negative pairs of words apart from each other. Finally, Section 5 gives a summary of the chapter and reaches a number of conclusions.

5.1 Introduction

Language is an important means of communication. In the course of communication, three fundamental elements come into play. These elements are thought of as standing in a quite definite relationship. The first element is the *message*, an utterance in speech or writing that contains information about something. The second element is the *meaning*, the idea that a language user intends to express and convey. Expressing meaning is what language is all about. Everything in language, grammatical, lexical or phonological constructions, contributes to express meaning. Meaning is a mental entity or some kind of abstract concept that is carried by an utterance. It is located in the mind and fused into the linguistic signs as it is produced. The third element is the *interpretation*, the act of understanding the meaning of an utterance. It is the act of determining what the meaning of an utterance is. This is the process by which a listener or reader uses his or her cognitive abilities and employs all of the available clues to understand an utterance. To understand how communication works, we need to understand how the meaning of an utterance is interpreted and to identify what principles account for it.

5.2 Theories of interpretation

The interpretation of utterance meaning has attracted some attention in the literature. In contemporary semantics, two theories exist. One is the reference theory, which regards meaning as objective in nature. Lexical meaning is a matter of relationships holding between words and mind-independent worlds. To count two lexemes as non-synonymous, the choice between them relies on a difference in the external world represented by sentences containing them. Truth-conditional theory is preoccupied with how language relates to reality. The meaning of an utterance can be specified by giving the condition under which it is true. Another is the conceptualist theory, which considers meaning as subjective in nature. Meanings tend, as Traugott (1986: 540) claims, to refer less to objective situations and more to subjective ones; less to the described situation and more to the discourse situation. Construal theory is concerned with how human beings use language to represent, or even construct, reality. The meaning of an utterance represents the way the speaker describes a situation. In what follows, I briefly survey these two approaches to interpretation which are prevalent in modern linguistics.

5.2.1 Reference theory

The reference theory of meaning is an approach to language in which the meaning of an utterance equals its reference. The meaning of an utterance resides in the relationship between the utterance and aspects of an objective world. This theory focuses exclusively on the referential properties of language. Language is seen as corresponding to the external world in an almost literal sense. The meaning of an utterance involves applying it appropriately to an object in the world. This theory has given rise to a more sophisticated version, referred to as truth-conditional theory. Truth-conditional theory or Formal semantics is an approach to language that sees the meaning of an utterance as the same as, or reducible to, its truth conditions. The meaning of an utterance is identified with the conditions in the world under which it is true or false. Speakers of language are concerned with the conditions which allow them to determine the objective truth, or otherwise, of utterances. At issue here is the logical basis of utterances. Examining the logical foundation of language entails employing the metalanguage of formal logic in the description of utterances. For example, knowing what a sentence like *The wind is blowing today* means involves knowing what situation in the world this would correspond to. On the basis of such knowledge, speakers can judge the statement to be either true or false.

Although this theory has been at the heart of work by many linguists, it does not offer a complete account of utterance meaning. Matching up a given expression with a certain kind of situation is a valid technique but it is not the whole story. There is more to the meaning of an utterance than the relationship between the utterance and its referents. One obvious limitation, stressed by Taylor (2002: 188–9), is that the theory is applicable only to expressions which designate concrete entities. No speaker can point to entities named by words like *beauty*, *kind*, *fair*, for example. Another limitation, cited in Bussman (1996: 498), is that the theory fails in the case of expressions which refer to the same thing, which it assumes to be synonymous. For example, *non-person* and *unperson* refer to the same thing in the world, but they differ in meaning. The first refers to one who is ignored. The second refers to one who is non-existent. A further limitation, detected by Goddard (1998: 5), is that the property of making reference does not belong to the utterances themselves, but rather to their uses on particular occasions. Words like *I, now, this* and *here* can refer to many things depending on context. Finally, not all utterances have truth conditions. Utterance types such as interrogatives or imperatives are, unlike declaratives, neither true nor false. Owing to such limitations, reference theory is no longer a solution to the problem posed by negative prefixes. Such a state of affairs paves the way for a new theory of meaning.

5.2.2 Conceptualist theory

The conceptualist theory of meaning is an approach to language that describes the meaning of an expression as the idea in the mind of the person who uses it. Language is seen as a kind of concept located within users' conscious experience. The meaning of an utterance is described in terms of the way it is conceptualised. In communication, the production of an utterance undergoes, I propose, two stages. The two stages operate concurrently. The first stage has to do with conceptualisation, where the speaker makes the decisions about how to frame an idea to language. At this conceptual stage, the speaker forms a concept in the mind about a particular situation. The second stage has to do with formulation, where the speaker frames the message into words, phrases and clauses. At this linguistic stage, the speaker chooses the lexical resources available in language to stand for his or her communicative intent.

All this shows that speech does not start from nothing. It starts from concepts. Both conceptualisation and formulation are tied together in producing any utterance. For example, when the speaker wants to describe a situation, s/he first generates a picture of it in the mind and second seeks in the mental lexicon the words that best represent it.

A modern version of this position is held by linguists like Lakoff and Langacker and is referred to as Cognitive Semantics, an approach to language that equates the meaning of an utterance with its conceptualisation. The meaning of an utterance, as shown by a variety of papers edited in Taylor & MacLaury (1995) and Goldberg (1996), is not defined in terms of the conditions in the real world under which it can be judged as true or false. Rather, it is defined in terms of the use of the utterance in communication, which is the result of the speaker's construal of a situation. Knowing the meaning of an utterance involves more than simply being able to identify it. Underlying this theory is the idea that linguistic expressions do not refer directly to the world. Rather, they refer to entities in a mental space. The mental space could be both objective, referring to things that really exist in the world, and non-objective, referring to things that do not link up with things in the real world. As we have seen, the utterances *non-person* and *unperson* are truth-conditionally equivalent, i.e. they describe the same objective situation, but they may nevertheless have different meanings. Therefore, linking utterances to situations or observable features is not enough for semantic differences. The utterances differ with respect to how they construe the described situation. Construal is then a matter of how a situation is conceptualised and how it is linguistically encoded.

5.3 Negatively-prefixed constructions

In the present analysis, I extend the *construal* theory to morphology. I attempt to show how the concept of construal is relevant to the interpretation of an utterance containing a negative prefix. On a general basis, I follow Langacker (1991: 338) who writes: 'grammatical structure reduces to patterns for the structuring and symbolisation of conceptual content, and that all valid grammatical constructs have some kind of conceptual import'. Grammar is symbolic in nature. It is inherently meaningful, not autonomous or accidental. This implies that morphological patterns in language are semantically motivated. Applying this to morphology, I argue that linguistic elements such as prefixes are viewed as meaningful in all their uses. Their meanings shape the morphological structures of the words in which they appear. The choice of a prefix is based on the way the speaker conceptualises a situation. Its use in discourse is a direct response to the needs of communication.

On a specific basis, I build on my own work (Hamawand, 2007). A complex utterance consists of a base and a prefix. The two component substructures are argued to have conceptual content. The base has multiple contents. The prefix has its own content which it adds to that of the base. Of the two component substructures, the prefix is more important because it is responsible for the semantic make-up of the derived word. More precisely, the prefix serves to highlight a particular facet within the content of the base. A complex word consisting of a base plus a prefix has, thus, a new meaning. The use of the prefix is the result of construal, the particular image the speaker selects to structure the content of an expression. It is the image that the speaker imposes on the conceptual content of a base relative to the communicative needs of the discourse. Because the base is multi-faceted, the speaker has the chance to construe it alternatively and use different prefixes to represent the construals.

5.3.1 Assumptions

To carry out the analysis here, this study makes use of two assumptions that are fashionable in modern linguistics:

1) A difference in syntactic form always indicates a difference in meaning, according to Bolinger (1977). Lexical and syntactic variation is not free; variants usually display subtle differences in meaning or unequal functions in discourse, which can be observed in certain contexts. Appealing to both intuition and empirical evidence, he shows that a difference in form of two words or sentences entails a difference in meaning. Clark (1987: 2) follows Bolinger's footsteps and proposes the *Principle of Contrast*: 'Every two forms contrast in meaning'. Furthermore, Clark claims (1987: 28) that 'The principle of contrast is inherent in language'. In this study, the aim is to show that the speakers of a language do not accept two terms as true synonyms. Whenever the language provides two words that appear alike on the face of it, the speakers should try to find a way to discriminate between them. Following Bolinger's (1977) dictum that each lexical item corresponds to a separate meaning, I argue that the difference in meaning is due to the presence of the prefixes.

2) Grammatical structures reflect aspects of imagery inherent in image-schemas, i.e. pre-linguistic conceptions, which are grounded in everyday physical or bodily experience, according to Lakoff (1987) and Johnson (1987). According to Langacker (1997: 4–5), 'A semantic structure includes both conceptual content and a particular way of construing that content'. Two expressions may invoke the

same conceptual content, yet differ semantically by virtue of the construals they represent. The ability of the speaker to construe an objective situation in different ways is fundamental to grammatical organisation. In this study, the aim is to show that lexical utterances reflect the encoding of conventional imagery. Each lexical utterance is assumed to be a conventional unit in a speaker's lexicon and to be semantically distinct from the other utterances. In Wierzbicka's (1988: 14) words, 'surface differences point to differences in meaning. Differences in grammatical form are not arbitrary, but signal differences in meaning'.

5.3.2 Goals

To realise the assumptions, the study sets itself the following three important goals:

1) Confirming the absence of synonymy in language, a phenomenon whereby word pairs are assumed to have similar meanings. It is very rare that words are randomly interchangeable. Although word pairs may share some features, they are still distinguishable in actual use. Word pairs may be truth-conditionally similar, but they are subtly different. Such is the nature of language that there is essentially some difference. In this work, the differences between pairs of negative words are often encoded morphologically through the use of different prefixes. Each prefix affects the character of the word by adding a special meaning to it. English is a language in which the morphological properties of words reflect subtle differences.

2) Verifying the multiplicity of word meanings in language, a phenomenon whereby a word base has a wide range of meanings. Meanings are not fixed entities, but rather constitute different subsets of semantic components. Meaning represents a highly motivated process, whereby word pairs can be derived from the same bases. In this work, one and the same state of affairs, indicated by a word base, can be linguistically encoded in different ways by means of different prefixes. Although the word pairs are based on the same content or designate the same entity, each has a different meaning, and so hosts a different morpheme. The word pairs acquire special meanings, which differ with respect to how they construe a given situation.

3) Substantiating the role of construal in language, a phenomenon whereby a speaker conceives of and expresses a situation. The speaker

is endowed with the ability to construe a situation in different ways, and choose the language resources available to express them. In this work, the differences between word pairs can be explained by the different construals imposed on their common content. The specific form of a word reflects the particular way in which the speaker chooses to describe its scene. Seemingly similar words are not free variants. The distinction between them is not a function of formal rules; it is exclusively a property of construal, which gives the speaker the flexibility to construe them in alternate ways.

5.3.3 Procedures

To achieve its goals, this study adopts the following three important steps:

1) Collecting the word pairs that begin with different negative prefixes but share the same bases. Given the lack of appropriate software tools, I conduct a manual search of word pairs by putting the lists of words containing the negative prefixes side by side in order to pick out the pairs. The use of a pair has double import. Theoretically, it achieves emphasis by placing focus on a particular segment within a word, and provides evidence that the segments compared have different meanings. Empirically, it helps, by relying on a corpus or texts from the Internet, to determine the contextual preferences of the pair members, and to stress the role of the rival prefixes in signalling the meaning differences between them.
2) Distinguishing the word pairs by identifying their individual behaviour. To achieve this, I first define the bases from which the pairs come, and second define the new derivatives to show how they differ from the base. Then, I provide examples of sentences to demonstrate their uses. The sentences are drawn from the British National Corpus and major online English dictionaries. To make the sentences reader-friendly but still rigorous in effect, I have shortened them by deleting non-essential elements from them. Through these examples, it becomes easy to see how rival prefixes serve as a locus of meaning difference in meaning, how related words are used in different ways, and how they are appropriate in different contexts.
3) Identifying the construals that are responsible for disambiguating the meanings of the word pairs. One great advantage of construal is that it captures differences between apparently similar words. Construal serves to show how the conceptual content of a base is viewed and

what difference the conceptual content of a prefix adds to the overall meaning of the derivation. As evidence, I make use of the technique of *collocation*, the tendency of certain words to occur together in a text. The information provided by the collocation analysis can be used as a major source of evidence for the allocation of a specific meaning to an occurrence of a word within a stretch of text, removing thus the potential ambiguity surrounding its use.

5.4 Operative construals

In this section, I deal with the construals that have a special effect in shaping the speaker's messages. This section discusses two types of construal: first those which operate within the same domains, and second those which operate between different domains.

5.4.1 Intra-domain construals

Intra-domain construals are responsible for interpreting rivalry between prefixes within individual domains. They account for cases when a word pair is derived from the same base by means of two or more prefixes belonging to the same domain. The construals serve to show that even if word pairs are related by derivation, their meanings are quite different. Each member of a pair symbolises a specific construal which provides a solution to its interpretation. It has its own sense, which is conceptually sparked by the particular construal imposed on its base content and linguistically encoded by the use of the appropriate prefix. Each prefix focuses on a particular facet of the content and contributes to the overall meaning of the derivation. An understanding of construal, then, makes a distinguishable difference in how the word pairs are used. In what follows, I present each of the domains and explain the role of construal in disambiguating the meanings of a range of word pairs.

5.4.1.1 The domain of distinction

Negative prefixes that can act as rivals within the domain of *distinction* are *a-*, *dis-*, *in-*, *non-* and *un-*. Nevertheless, each prefix represents a different construal of the base, and so has a meaning of its own. Following Zimmer (1964: 33) and Horn (2001: 280), I argue that the prefixes deliver either a contradictory meaning or a contrary meaning, depending on the meaning of their base. Relative to the non-gradable nature of the combining bases, *non-* and (to a lesser degree) *a-* yield a contradictory reading, and so trigger an objective description. *Non-* represents the highest degree of contradictoriness. It is distinctly impartial in tone. *A-* represents the lowest degree of contradictoriness. It is less impartial in tone. Relative to the gradable nature of the combining bases, *dis-*, *un-* and *in-* yield a contrary reading, and so produce a subjective evaluation. These are partial in flavour. *Dis-* represents the lowest degree of contrariety. It is unfavourable in character. Right in the middle is *un-*, which is disapproving in character. At the highest degree of contrariety is *in-*. which is critical in character.

5.4.1.1.1 non- vs. a-

The negative prefixes *non-* and *a-* evoke the domain of *distinction*, but they represent different facets of it. The prefix *non-* means 'different from the quality described by the adjectival base'. It serves to show that the entity described is by no means related to the thing specified by the base. In this function, it is merely descriptive; hence the word it derives acquires a contradictory reading. By contrast, the prefix *a-* means 'divergent from the quality referred to by the adjectival base'. It serves to show that the entity described is related to the thing specified by the base, but it is not willing to do it or have it. In this function, it is evaluative; hence the word it derives obtains a contrary reading. Of the two prefixes, *non-* is stronger in expressing distinction than *a-*. A trawl through the data in the corpus and the Internet leads to the following key remarks. Words beginning with *non-* apply to areas of knowledge. Words beginning with *a-* apply to animates, including people and animals. By way of illustration, consider the following negative pair:

(1) non-social vs. asocial[1]

a. Questions have been raised about the *non-social* areas of linguistics.

b. They refused him because he was solitary, *asocial* and uncooperative.

The two adjectives are derived from the base *social*, meaning 'relating to human society, or involving pleasant companionship'. Nevertheless, each adjective represents a different construal of the base, and so has its own definition. In (1a), the adjective *non-social* means 'not socially oriented'. *A non-social area* is one that lacks a social component. *Non-social* collocates with words referring to fields of knowledge such as *area, aspect, discipline, resource, subject*, etc. In (1b), the adjective *asocial* means 'unwilling to mix socially'. *An asocial person* is one who is averse to social interaction. *Asocial* collocates with words referring to people or animals such as *bear, cat, man, people, person*, or their deeds such as *approach, behaviour, conduct, manner, intention*, etc. Of the two adjectives, only *asocial* accepts adverbs of degree, as in *He is totally asocial.*[2]

5.4.1.1.2 *non-* vs. *dis-*

The negative prefixes *non-* and *dis-* evoke the domain of *distinction*, but they embody different facets of it. The prefix *non-* means 'different from the quality described by the adjectival base'. It expresses neutral negation, showing that the entity described fails to perform the action specified by the base, probably for reasons within its control. It simply forms an opposition without providing any details. By contrast, the prefix *dis-* means 'the converse of the quality signified by the adjectival base'. It expresses evaluative negation, showing that the entity described fails to perform the action specified by the base, probably for reasons beyond its own control. Not only does it form an opposition, but it also provides some details. A search of the data in the corpus and the Internet leads us to the following observations. Words taking *non-* apply to humans. Words taking *dis-* apply to humans and things. Consider, for instance, the following negative pair:

(2) non-appearance vs. disappearance
   a. They were probably very disappointed at the band's *non-appearance.*
   b. A man was being questioned in connection with their *disappearance.*

The two nouns are derived from the base *appearance*, meaning 'to become noticeable or to be present'. Nevertheless, construal connects them to different uses. In (2a), the noun *non-appearance* means 'failure to appear or attend'. *A band's non-appearance* is its failure to appear in a concert. *Non-appearance* collocates with words referring to people such as *band, ensemble, fans, squad, worker*, etc. In (2b), the noun *disappearance* means 'to pass out of existence or notice'. Somebody's disappearance is the act of disappearing or satate of having vanished. *Disappearance* collocates with words referring to people such as people such as *founder, predator, veteran, witness, youth*, or things such as *border, elm, jewel, money, sign*, etc. Of the two nouns, only *disappearance* accepts adjectives implying activity, as in *The sudden disappearance of the man provoked anger.*[3]

5.4.1.1.3 *non-* vs. *un-*

The negative prefixes *non-* and *un-* share the domain of *distinction*, but they symbolise distinct facets of it. The prefix *non-* means 'different from the quality described by the adjectival base'. It underlines a comparison in which one entity is described as being quite different from another. In this use, it is objective, used to stress the other nature of an entity. For its part, the prefix *un-* means 'the antithesis of what is specified by the adjectival base'. It underscores a comparison in which one entity is described as being related to another, but it tends to breach it. In this use, it is evaluative, used to criticise an entity. A perusal of the data in the corpus and the Internet produces the following main points. Words beginning with *non-* describe people, placing emphasis merely on the denial of the content signified by the bases. Words beginning with *un-* describe people's behaviour, laying emphasis on the evaluation of the content signified by the bases. Consider, by way of illustration, the following negative pair:

(3) non-professional vs. unprofessional
   a. The company provides training for the *non-professional* staff.
   b. She was found guilty of *unprofessional* conduct by the court.

The two adjectives are derived from the base *professional*, meaning 'relating to a profession, or conforming to its standards'. Nonetheless, each adjective imposes a different construal on the base, and so has its own meaning. In (3a), the adjective *non-professional* means 'not trained in a specific profession'. *Non-professional staff* are staff who are not professionals. *Non-professional* collocates with words referring to personnel such as *actors, drivers, helpers, staff, workers*, etc. In (3b), the adjective *unprofessional* means 'below or contrary to the standards expected in a particular profession'. *Unprofessional conduct* means conduct that does not reach the standard expected in a particular profession. *Unprofessional* collocates with words referring to demeanour such as *behaviour, conduct, deportment, manner, practice*, etc. Most often, the adjective occurs in *it*-constructions, denoting a temporary property, as in *It is unprofessional to argue the case publicly*. As evidence, only *un-* has negative connotation. Using Algeo's (1971: 90–1) test, *a non-professional staff* can still be competent, whereas *an unprofessional staff* is not.[4]

5.4.1.1.4 non- vs. in-

The negative prefixes *non-* and *in-* point to the domain of *distinction*, but they emphasise distinct facets of it. The prefix *non-* means 'different from the quality described by the adjectival base'. In comparing two entities, it refers to the mere description of a state of affairs. It triggers a description that is free of any bias or judgement. By contrast, the prefix *in-* means 'the opposite of the property expressed by the adjectival base'. In comparing two entities, it refers to the evaluation of a state of affairs. It suggests an evaluation that amounts to disapproval, pointing out one or more faults with the entity under assessment. Scanning through the data in the corpus and the Internet leads us to the following findings. Words beginning with *non-* identify the nature of their companions which include concrete nouns like items or people. Words beginning with *in-* identify values of their companions which include abstract nouns like features or properties. By way of example, examine the following negative pair:

(4) non-essential vs. inessential

a. A great proportion of their income is spent on *non-essential* goods.

b. Most women work part-time because their earnings are *inessential*.

The two adjectives are derived from the base *essential*, meaning 'fundamental, necessary, indispensable'. However, each adjective signifies a different construal of the base, and so has a distinct use. In (4a), the adjective *non-essential* means 'not necessary'. *Non-essential goods* are goods that are simply not fundamental. *Non-essential* collocates with words referring to items such as *articles, goods, products, stuff, wares*; or personnel such as *employees, labourers, staff, users, workers*, etc. In (4b), the adjective *inessential* means 'not necessary'. *Inessential earnings* are earnings that are good but not important. *Inessential* collocates with words referring to abstract entities such as *concepts, details, earnings, features, properties*, etc. Of the two adjectives, only *inessential* accepts adverbs of degree as in *The earnings are extremely inessential.* This analysis agrees with Algeo's (1971: 90–1) findings, according to which one can say *The civil law is non-religious* (not religious), but one cannot say *The civil law is irreligious* (hostile to religion).[5]

5.4.1.1.5 *a-* vs. *un-*

The negative prefixes *a-* and *un-* deal with the domain of *distinction*, but they incarnate different facets of it. The prefix *a-* means 'divergent from the quality referred to by the adjectival base'. In the process of negation, it indicates that the entity described in the base is irregular. That is, it is bad but normal. The prefix *un-* means 'the antithesis of what is specified by the adjectival base'. In the process of negation, it indicates that the entity described in the base has deviant properties. That is, it is both bad and odd. Skimming through the data in the corpus and the Internet reveals the following facts. Words beginning with *a-* expound on features of things, which are neutral in effect. Words beginning with *un-* impart on properties of processes, which are harmful in effect. Examine, by way of example, the following negative pair:

(5) atypical vs. untypical

a. There is a need to develop childcare at *atypical* times.

b. Burglary, for example, is an *untypical* female crime.

The two adjectives are derived from the base *typical*, meaning 'having the distinctive qualities of a particular type'. Yet, each adjective incarnates a different construal, and so has its individual sense. In (5a), the adjective *atypical* means 'not usual'. *Atypical time* is time that does not conform to the usual type or expected pattern. For example, it may be outside 8am to 6pm. *Atypical* collocates with words referring to happenings such as *conditions, situations, states, times, working hours*, or signs such as *cells, features, forms, moles, symptoms*, etc. In (5b), the adjective *untypical* means 'not characteristic'. *Untypical crime* is crime that is not representative of a group, class, or type. *Untypical* collocates with words implying action such as *act, application, crime, event, gesture*, etc. Of the two adjectives, only *untypical* can be modified by adverbs implying negativity, as in *He wrote an utterly untypical work.*[6]

5.4.1.1.6 a- vs. in-

The negative prefixes *a-* and *in-* deal with the domain of *distinction*, but they embody divergent facets of it. The prefix *a-* means 'divergent from the quality referred to by the adjectival base'. It portrays an entity as being just the opposite of the content denoted in the base. The entity does not care about what the content of the base denotes, and thus remains outside its scope. The prefix *in-* means 'the opposite of the property expressed by the adjectival base'. It portrays an entity as violating or infringing the content denoted in the base. The entity does things that are contrary to what the content of the base denotes, and thus is in conflict with its scope. Checking through the data in the corpus and the Internet leads to the following observations. Words beginning with *a-* comment on people or their attitudes. Words beginning with *in-* comment on people or their acts. To explicate the contrast, we need to examine a negative pair:

(6) amoral vs. immoral
  a. He is *amoral*, paranoid and devoid of convictions.
  b. He condemned the government's action as *immoral*.

The two adjectives are derived from the base *moral*, meaning 'behaving in ways considered by most people to be correct and honest'. Even so, each adjective embodies a different construal, and so is distinctive in use. In (6a), the adjective *amoral* means 'without moral principles'. *An amoral person* is a person who has no moral sense; who is unconcerned with right or wrong. *Amoral* collocates with words referring to people such as *dealer, humans, instigator, practitioner, scientist*, or their attitudes such as *cunning, fascination, feeling, indifference, nature*, etc. In (6b), the adjective *immoral* means 'morally wrong'. *An immoral act* is one that is not in conformity with the moral code of behaviour. *Immoral* collocates with words referring to people such as *government, leader, man, society, woman*, or their acts such as *action, behaviour, conduct, deed, practice*, etc.[7]

5.4.1.1.7 *dis-* vs. *un-*

The negative prefixes *dis-* and *un-* elicit the domain of *distinction*, but they exemplify different facets of it. The prefix *dis-* means 'the converse of the quality signified by the adjectival base'. It shows distinction in medial degree, i.e. situated in or towards the middle on the scale of contrast. It draws attention to only the necessary information in the contrast. By contrast, the prefix *un-* means 'the antithesis of what is specified by the adjectival base'. It shows distinction above medial degree, i.e. situated between the middle and the top on the scale of contrast. It sheds light on some of the information in the contrast. Scrutiny of the data in the corpus and the Internet shows that words beginning with *dis-* collocate mostly with nouns referring to people or their acts, whereas words beginning with *un-* collocate mostly with nouns referring to people or things. To elucidate the contrast, we need to examine a negative pair:

(7) disinterested vs. uninterested

   a. They need the help of some *disinterested* historians.

   b. They are completely *uninterested* in such questions.

The two adjectives are derived from the base *interested*, meaning 'wanting to give one's attention to something and discover more about it'. Despite that, each adjective manifests a different construal, and so plays a different role. In (7a), the adjective *disinterested* means 'impartial', having no personal involvement or receiving no personal advantage, and therefore free to act fairly. *A disinterested historian* is one who is free from bias or self-interest. *Disinterested* collocates with words referring to people such as *assessor, observer, historian, investigator, researcher*, or their acts such as *argument, judgement, research, scholarship, speculation*, etc. In (7b), the adjective *uninterested* means 'indifferent'. *Uninterested people* are people who show no interest in something or take no heed of it. *Uninterested* collocates with words referring to people such as *child, lady, mother, public, student*, or things such as *conversation, game, idea, problem, question*, etc. Of the two adjectives, only *disinterested* can co-occur with positive words, as in *He is disinterested and honourable in his public life.*[8]

5.4.1.1.8 *un-* vs. *in-*

The negative prefixes *un-* and *in-* exemplify the domain of *distinction*, but they activate different facets of it. The prefix *un-* means 'the antithesis of what is specified by the adjectival base'. It underlies contrast that lies above the middle point on the scale. It concentrates on less detail in highlighting the contrast. It carries the message that the action profiled is in a state that is possible but difficult to do. By contrast, the prefix *in-* means 'the opposite of the property expressed by the adjectival base'. It underscores contrast that lies at the end of the scale. It shows distinction to an extreme degree, and draws on fuller detail in highlighting the contrast. It conveys the message that the action profiled is in a state that is impossible to do. Scrutiny of the data in the corpus and the Internet illustrates that words beginning with *un-* tend to take nouns denoting concrete things or tangible objects. By contrast, words beginning with *in-* tend to collocate with nouns denoting abstract things or intangible objects. To illuminate the contrast, we need to examine some negative pairs:

(8) unrepairable vs. irreparable
  a. They broke four legs off a bed making it *unrepairable.*
  b. The students did *irreparable* damage to the computers.

The two adjectives are derived from the base *repairable,* meaning 'abe to be repaired'. In spite of that, each adjective epitomises a different construal, and so has a different application. In (8a), the adjective *unrepairable* means 'impossible to repair'. *An unrepairable bed* is one that is in so bad a physical condition that it cannot be mended. *Unrepairable* collocates with words denoting objects such as *bed, camera, clock, table*, *watch*, etc. In (8b), the adjective *irreparable* means 'impossible to repair'. *Irreparable damage* is damage that is not likely to be reverted. *Irreparable* collocates with words denoting harm such as *blow, damage, harm, injury, loss*, etc. For this reason, only *irreparable* is possible in contexts like *Her death is an irreparable loss*. In such contexts, it is impossible to rectify, repair or restore the state of affairs designated by the base, and that is exactly the core meaning of the prefix.

(9) unmeasurable vs. immeasurable
  a. The consequences of a nuclear war are *unmeasurable.*
  b. Her films had an *immeasurable* impact on the watchers.

The two adjectives are derived from the base *measure*, meaning 'a unit used for stating the number, size or weight of something'. Despite that, each adjective stands for a different construal, and so is used in a different context. In (9a), *unmeasurable* means 'too awkward to measure'. *Unmeasurable consequences* are consequences that cannot be calculated by any known measure or scale. *Unmeasurable* collocates with words implying boundedness such as *changes, consequences, effects, features, values*, etc. In (9b), *immeasurable* means 'too vast to measure'. *An immeasurable impact* is an impact that is too large or great that it cannot be known exactly. *Immeasurable* collocates with words implying unboundedness such as *experience*, *impact, importance, significance, value*, or emotions such as *harm, joy, passion, sorrow, suffering*, etc. As evidence, one can say *an immeasurable sense of regret*, but not *an unmeasurable sense of regret*. The reason is that regret is unbounded in nature.[9]

5.4.1.2 The domain of opposition: anti- vs. counter-

Negative prefixes that can act as rivals within the domain of *opposition* are *anti-* and *counter-*. Even so, they are not identical for they fulfil separate functions in the language. The negative prefix *anti-* means 'reacting against the thing named by the nominal base'. *Anti-* refers to the ideology adopted to respond to another ideology. By contrast, the negative prefix *counter-* means 'facing the action or activity designated by the nominal base'. It refers to an action that takes place in response to another action. In essence, *anti-* suggests a diplomatic and less confrontational line than the negative prefix *counter-*. An examination of the data in the corpus and the Internet shows that distinct patterns are discernible in the use of each prefix. Collocations of words negated by *anti-* refer to ordnances or statutes. By contrast, collocations of words negated by *counter-* refer to people and their workplaces. This contrast is borne out by negative pairs such as the following:

(10) anti-terrorism vs. counter-terrorism

a. Those arrested under *anti-terrorism* laws were eventually released.

b. The crowd watched the movements of the *counter-terrorism* team.

The two nouns are derived from the base *terrorism,* meaning 'the use of violence, torture, or physical intimidation by a group or organisation as a means of forcing others to satisfy its demands'. Although the nouns are similar in derivation, construal restricts them to separate uses. In (10a), the noun *anti-terrorism* refers to the attitude or ideology which opposes terrorism. *Anti-terrorism laws* are laws that are designed to fight terrorism. *Anti-terrorism* collocates with words referring to laws and legislation. In (10b), the noun *counter-terrorism* refers to the practices or tactics which oppose terrorism. *A counter-terrorism team* is a team that specialises in fighting terrorism. *Counter-terrorism* collocates with words referring to people such as *experts, forces, specialists, teams, units*, or the places where they work such as *police headquarters, inspectorates, intelligence agencies, security services*, etc. As evidence, only *counter-* is modified by adverbs denoting action which is due to the nature of the verbal base it combines with, as in *The police swiftly counter-acted to tackle the problem.*[10]

(11) anti-revolution vs. counter-revolution

a. The *anti-revolution* bill limits the activity of democratic forces.

b. A *counter-revolution* has occurred, without a shot being fired.

The two nouns are derived from the base *revolution,* meaning 'a change in the way a country is governed, usually to a different political system and often using violence or war'. Although the nouns are similar in derivation, construal restricts them to separate uses. In (11a), the noun *anti-revolution* refers to an opinion or general feeling which opposes revolution. An *anti-revolution bill* is a piece of legislation that is designed to fight revolution. *Anti-revolution* collocates with words referring to laws and legislation and verbs denoting attitude such as *express, hold, reflect, state, voice*, etc. In (11b), the noun *counter-revolution* refers to a revolution that is intended to replace a previous revolution. *Counter-revolution* is a violent action against a government which came to power as a result of a revolution in order to destroy or replace it. *Counter-revolution* collocates with verbs denoting occurrence such as *happen, occur, start, take place, wage*, etc. As evidence, one can say *wage a counter-revolution* but not *anti-revolution*, because *anti-revolution* is not a revolution, and so cannot be waged.

5.4.1.3 The domain of privation: dis- vs. un-

Negative prefixes that can act as rivals within the domain of *privation* are *dis-* and *un-*. Nonetheless, each prefix evokes a different construal, and so has a distinct role to play in the language. The negative prefix *dis-* means 'lacking the thing signified by the nominal base'. It picks out the action that is implied by the base. In contrast, the negative prefix *un-* means 'bereft of what is specified by the nominal base'. It picks out the state that is implied by the base. In essence, *dis-* is less powerful than *un-* in expressing negation. An examination of the data in the corpus and the Internet shows that there are distinct patterns associated with each prefix. Words prefixed by *dis-* tend to collocate with expressions of disapproval of statements, or surprise at events. By contrast, words prefixed by *un-* tend to collocate with expressions of lack of faith in a system or scepticism of a doctrine. This contrast is illustrated by negative pairs such as the following:

(12) disbelief vs. unbelief

  a. They listened in *disbelief* to the extraordinary story.
  b. The essence of *unbelief* lies in departing from God.

The two nouns are derived from the base *belief*, meaning 'the feeling of certainty that something exists or is true'. Even though the nouns stem from the same base, construal confines them to distinct uses. In (12a), the noun *disbelief* means 'lack of belief'. *Disbelief* implies refusal to accept that an account or a story is true or real. *Disbelief* collocates with expressions denoting surprise or shock such as *ask/breathe/gasp/hoot/listen/stare/watch in...; give a snort of..., give way to..., shake the head in..., wear a look of...*, etc. In (12b), the noun *unbelief* means 'lack of faith'. *Unbelief* implies reluctance to believe that a faith or a religion is true. *Unbelief* collocates with expressions denoting attitude or principle such as *fill the heart up with ..., manifest a strong/weak...in, reveal the ...in*, etc. Evidence for this interpretation is provided by the fact that in a context where religion is in focus only the use of *unbelief* is permissible.

(13) disburden vs. unburden

a. She has not decided to *disburden* herself of the task.

b. He will *unburden* himself of the guilt for his actions.

The two verbs are derived from the base *burden*, meaning 'something difficult or unpleasant that you have to deal with or worry about'. Even though the nouns stem from the same base, construal confines them to distinct uses. In (13a), the verb *disburden* means 'to relieve of a burden or responsibility'. *To disburden oneself* is to free oneself from or cast out something objectionable or undesirable. *Disburden* collocates with words referring to commitments such as *charge, duty, obligation, responsibility, task*, etc. In (13b), the verb *unburden* means 'to get something off your mind'. *To unburden oneself* is to relieve oneself of something that has been worrying one by telling somebody about it so that one feels less conscious of it. *Unburden* collocates with words referring to emotions such as *concern, fear, guilt, regret, worry*, etc. As can be seen, the two verbs are typically followed by reflexive pronouns.

5.4.1.4 The domain of removal: de- vs. un-

Negative prefixes that can act as rivals within the domain of *removal* are *de-* and *un-*. Nevertheless, they are distinctive in use, which is due to the way the base is construed. The prefix *de-* means 'removing the thing described by the nominal base'. Prototypically, it focuses on the removal of non-humans, but exceptionally it targets a human. The prefix *un-* means 'taking away what is specified by the nominal base'. It focuses on objects in the process of removal, denoting mainly separation. Browsing of the data in the corpus and the Internet yields the following observations. Word with *de-* prefer noun collocates which refer to humans. Words with *un-* prefer noun collocates which refer to objects. Consider this contrast in negative pairs such as the one below:

(14) decouple vs. uncouple

a. The revenues had been *decoupled* form the sales.

b. The carriages had been *uncoupled* from the train.

The two verbs are derived from the base *couple*, meaning 'to connect or combine two things'. In spite of that, each prefix has its own particular way of combining with the base. In (14a), the verb *decouple* means 'to disengage one thing from another'. *To decouple two things* is to end the relationship between them. *Decouple* collocates with words referring to disjoined objects such as *age from size, earnings from revenues, policy from ideology, revenues from sales, software from hardware*, etc. In (14b), the verb *uncouple* means 'to disconnect one thing from another'. *To uncouple two things* is to separate two things that are joined together, by undoing a fastening that connects them. *Uncouple* collocates with words referring to joined objects such as *carriage from train, hook from chain, ring from lock, sleeve from tube, trailer from tractor*, etc. As evidence, one can say *decouple foreign policy from ideology*, but one cannot say *uncouple foreign policy from ideology*.

(15) defrock vs. unfrock
   a. The bishop had the authority to *defrock* the priest.
   b. They used the song to *unfrock* the front-row girls.

The two verbs are derived from the base *frock*, nowadays meaning 'a girl's dress'. In spite of that, each prefix combines in its own way with the base. In (15a), the verb *defrock* means 'to remove somebody from a position of honour or privilege'. *To defrock a priest* is to formally remove him, as a punishment for wrongdoing. *Defrock* collocates with words referring to people such as *Cardinal, minister, monk, priest, theologian*, etc. In (15b), the verb *unfrock* means 'to remove a frock from somebody'. To unfrock a person is to remove his or her clothes. *Unfrock* collocates with words referring to people such as *baby, child, doll, girl, kid*, etc. As evidence, in a context where a concrete entity, e.g. clothes or ornaments, receives emphasis only the use of *unfrock* is admissible in discourse.

5.4.1.5 The domain of reversal: de- vs. un-

Negative prefixes that can act as rivals within the domain of *reversal* are *de-* and *un-*. Despite having the same base, construal assigns them discrete roles. The prefix *de-* means 'reversing the action described by the nominal base'. It applies to people by shutting them out or excluding them from a place or condition. The prefix *un-* means 'inverting what is specified by the verbal bases'. It applies to objects by opening or unlocking them. A study of the data in the corpus and the Internet suggests the following remarks. Words negated by *de-* collocate with nouns referring to humans. Words negated by *un-* collocate with nouns referring to objects. Examine this contrast in a negative pair such as the one below:

(16) debar vs. unbar
   a. They were *debarred* from the club for unacceptable behaviour.
   b. After hours of trying, they finally managed to *unbar* the door.

The two verbs are derived from the base bar, meaning 'to prevent someone from doing something or fasten something with a bar'. Despite that, construal imposes a certain distance between them in use. In (16a), the verb *debar* means 'to keep somebody from entering or taking part in something'. *To debar* a person is to prohibit him or her officially from doing something by law or official agreement: *debar* collocates with words referring to people such as *citizens, employees, seamen, students, travellers*, etc. It is to be noted that in *debar* the prefix functions as an intensifier. In (16b), the verb *unbar* means 'to unlock or open a door or gate'. *To unbar a door* is to open it by withdrawing a bolt. *Unbar* collocates with words referring to things such as *device, door, gate, handset, phone*, etc. As evidence, in a context like *preventing someone from voting in the elections*, only *debar* can replace *prevent*.

(17) devalue vs. disvalue

a. The country was forced to *devalue* its currency.

b. He *disvalued* the achievement we'd made so far.

The two verbs are derived from the base *value*, meaning 'to estimate the value of something'. Despite that, construal imposes a certain distance between them in use. In (17a), the verb *devalue* means 'to reduce the value of something, particularly money'. *To devalue something* is to lower the worth or importance of something. *Devalue* collocates with words referring to money such as *cash, currency, dollar, pound, sterling*, etc. In (17b), the verb *disvalue* means 'to reduce the regard of something or someone'. *To disvalue something or someone* is to treat something or someone as having little value, i.e. to think that something or someone is unimportant or ordinary and not give them the respect that they deserve. *Disvalue* collocates with words referring to people or their deeds such as *achievement, experience, expertise, qualification, sacrifice*, etc. That is why in showing disregard for someone's performance only the negative word *disvalue* is appropriate.

5.4.1.6 The domain of treatment: mis- vs. mal-

Negative prefixes that can act as rivals within the domain of *treatment* are *mis-* and *mal-*. However, they are used differently depending on the construal of the base. The prefix *mis-* means 'falsely performing the action conveyed by the verbal base'. It sheds light on the accidental or negligent nature of the action conveyed by the base. By contrast, the prefix *mal-* means 'inappropriately executing the action denoted by the verbal base'. It sheds light on the intentional or purposive nature of the action conveyed by the base. Whereas *mis-* causes psychological damage, *mal-* causes physical damage. A careful examination of the data in the corpus and the Internet shows that the behaviour of words hosting the two prefixes is patterned in observable ways. Words hosting *mis-* are found with noun collocates that often refer to things, and occasionally to humans or pets. By contrast, words hosting *mal-* are found with noun collocates that often refer to humans. This contrast is illuminated by negative pairs such as the following:

(18) misadjustment vs. maladjustment
  a. *Misadjustment* of the winder will affect the watch performance.
  b. Deprivation at home can lead to *maladjustment* of the children.

The two nouns are derived from the base *adjust*, meaning 'to change slightly so as to achieve a desired result'. Despite having the same base, construal assigns them discrete roles. In (18a), the noun *misadjustment* means 'false adjustment'. *Misadjustment* is used when something does not fit or fits badly, when one adjusts the winder of a watch badly so that it shows the wrong time. *Misadjustment* collocates with words referring to devices such as *button, clock, knob, switch, winder*, etc. In (18b), the noun *maladjustment* means 'inappropriate adjustment'. *Maladjustment* is used when someone is unable to adapt properly to his/her environment, leading often to problems in behaviour. *Maladjustment* collocates with words referring to humans such as *adolescents, children, individuals, persons, teenagers*; modifiers such as *marital, physical, psychological, sexual, social*, etc. As can be seen, *misadjustment* is used in technical contexts and denotes errors, whereas *maladjustment* is used in social contexts and denotes gaffes.

(19) mistreat vs. maltreat
  a. The woman doesn't let anyone *mistreat* her girls.
  b. He is charged with conspiracy to *maltreat* detainees.

The two verbs are derived from the base *treat*, meaning 'to behave in a particular way towards somebody or something'. Despite having the same base, they have independent uses. In (19a), the verb *mistreat* means 'falsely performing the action conveyed by the base'. *Mistreating girls* implies treating them unkindly or unfairly. The treatment is haphazard in nature; it amounts to insult or offence. *Mistreat* collocates with words referring to humans such as *children, girls, people, women, workers*, or *pets*, etc. In (19b), the verb *maltreat* means 'inappropriately executing the action denoted by the base'. *Maltreating detainees* implies treating them cruelly or brutally. The treatment is wilful in nature; it amounts to violence and torture. *Maltreat* collocates with words referring to people such as *captives, detainees, prisoners, strangers, tourists*, etc. Of the two verbs, only with *maltreat* can be used with verbs denoting physical power, as in *They maltreated the monument and then destroyed it.*[11]

5.4.2 Inter-domain construals

Inter-domain construals are responsible for interpreting rivalry between prefixes from different domains. They account for cases when a word pair is derived from the same base by means of two or more prefixes belonging to different domains. This is quite normal in Cognitive Linguistics, as a semantic unit can be conceptualised against more than one domain, which explains how it is employed. As Taylor (2002: 197) stresses, domains do not constitute strictly separated configurations of knowledge. They overlap and interact in numerous ways. The construals serve to show that although word pairs share the same bases, they are distinctive in use. For each member of a pair, the speaker construes the base differently, and so uses it discriminately. Each member of a pair has a different use, which is a function of the construal imposed by the speaker on the conceptual content of both the base and the prefix. Each prefix evokes a different domain and so represents a different construal. In what follows, I introduce the domains evoked by the prefixes, and the construals used to disambiguate the meanings of word pairs.[12]

5.4.2.1 non- vs. anti-

The negative prefixes *non-* and *anti-* represent different domains. *Non-* belongs to the domain of *distinction.* It means 'different from the quality described by the adjectival base'. It implies absoluteness, judging something to the highest degree. *Anti-* belongs to the domain of *opposition.* It means 'reacting against the thing named by the nominal base'. It implies relativity, judging something in comparison with something else. An investigation of the data in the corpus and the Internet shows that words having *non-* prefer nouns referring to animals or abstract things. By contrast, words having *anti-* prefer nouns referring to beliefs or chemical substances. To comprehend the contrast, let us discuss a negative pair:

(20) non-human vs. anti-human

a. Environmental disasters threaten both human and *non-human* life.

b. The philosophy behind such a theory of economy is *anti-human.*

The two words are derived from the base *human*, meaning 'of, relating to, or characteristic of mankind'. Still, construal makes a clear distinction between them in use. In (20a), the adjective *non-human* means 'not human'. *Non-human life* is life that does not belong to the human race. *Non-human* collocates with words referring to animals such as *creatures, mammals, primates, species, living things*, or abstract things such as *forces, life, sources, wealth, world*, etc. In (20b), *anti-human* means 'being against humanity'. *Anti-human philosophy* is philosophy that is against or opposed to human beings or human values. *Anti-human* collocates with words referring to views such as *attitude, concept, perspective, philosophy, stance*, etc. In another meaning, *anti-human* collocates with words referring to chemical substances such as *globulin, lactoferrin, pesticides, proteins*, etc. As evidence, it is right to say *anti-human in attitude* but wrong to say *non-human in attitude*.[13]

5.4.2.2 *un-* vs. *de-*

The negative prefixes *un-* and *de-* represent different domains. *Un-* activates the domain of *distinction*. It means 'the antithesis of what is specified by the adjectival base'. *De-* activates the domain of *reversal*. It means 'reversing the action described by the nominal base'. An investigation of the data in the corpus and the Internet shows that words with *un-* collocate with nouns referring to positions, whereas words with *de-* collocate with nouns referring to people or their performances. As Andrews (1986: 229) writes, the contrast becomes clear with pairs that have participial forms. In forms derived by *un-*, the quality in question has never existed, whereas in forms derived by *de-* the quality in question is established prior to the given speech event. To appreciate the contrast, let us consider a negative pair:

(21) ungraded vs. degraded
  a. How can he go to the employer with an *ungraded* degree?
  b. Why do so many people have to get *degraded* all the time?

The two words are derived from the base *grade*, meaning 'a level of quality, size or importance'. Yet, construal draws a sharp demarcation between them in use. In (21a), the adjective *ungraded* means 'not graded'. *An ungraded degree* is one that is not identified according to grade. *Ungraded* collocates with words referring to positions or courses such as *degree, level, order, rank, standing*, etc. In (21b), the verb *degrade* means 'to lower in character or status'. *Degraded people* are people who are reduced in esteem, position or importance. *Degrade* collocates with words referring to people, or things related to them such as *culture, integrity, life, performance, philosophy*, etc. Of the two prefixes, only *de-* accepts adverbs of manner, as in *These chemicals quickly degrade into harmless compounds*.[14]

5.4.2.3 *un-* vs. *mis-*

The negative prefixes *un-* and *mis-* represent different domains. *Un-* elicits the domain of *distinction*. It means 'the antithesis of what is specified by the adjectival base'. *Mis-* elicits the domain of *treatment*. It means 'falsely doing the action conveyed by the verbal base'. A look at the data in the corpus and the Internet indicates that *un-* words accept nouns referring to people or their habits, whereas *mis-* words accept nouns referring to accidents or occurrences. With *un-*, the harm is inflicted by interior sources, whereas with *mis-* it is caused by exterior sources. In addition, *un-* focuses on an occasional thing, whereas *mis-* focuses on a more permanent one. To understand the contrast, let us look at a negative pair:

(22) unfortunate vs. misfortunate

a. He was *unfortunate* to lose in the final round of the race.
b. He wasn't a bad sort, but he was *misfortunate* with drink.

The two words are derived from the base *fortunate*, meaning 'lucky'. However, construal maintains a noticeable difference between them in use. In (22a), the adjective *unfortunate* means 'having bad luck'. *An unfortunate person* is one who does not enjoy good luck. *Unfortunate* collocates with words referring to people such as *individuals, drivers, patients, residents, victims*, or their deeds such as *attempt, choice, crash, error, incident*, etc. In (22b), the adjective *misfortunate* means 'deserving or inciting pity'. *A misfortunate person* is one who is hapless, miserable or piteous. *Misfortunate* collocates with words referring to people such as *children, lady, soldiers, travellers*, or occurrences such as *cartoons, disease, events, happenings, mishaps*, etc. For this reason, *an unfortunate crash or death* is far more likely to be attested than *a misfortunate crash or death.*[15]

5.4.2.4 *dis-* vs. *mis-*

The negative prefixes *dis-* and *mis-* represent different domains. *Dis-* belongs to the domain of *privation*. It means 'depriving of the thing described by the nominal base'. *Mis-* belongs to the domain of *treatment*. It means 'falsely doing the action conveyed by the verbal base'. A glance at the data in the corpus and the Internet demonstrates that words to which *dis-* is attached co-occur with nouns describing people, whereas words to which *mis-* is attached co-occur with nouns describing things like motives, systems or services. To grasp the contrast, let us consider a negative pair:

(23) distrust vs. mistrust

a. She has a profound *distrust* of door-to-door salesmen.
b. He has a deep *mistrust* of electronic banking services.

The two words are derived from the base *trust*, meaning 'the confidence placed in a person or thing'. Nonetheless, construal applies them to different contexts. In (23a), the noun *distrust* means 'deprived of trust'. *Having distrust of salesmen* is to have no confidence in them. *Distrust* collocates with words referring to people such as *foreigner, judge, neighbour, salesman, writer*; or things such as *politics, research, science, technology, theory*, etc. In (23b), the noun *mistrust* means 'having no trust in'. *Having mistrust of something* is to have suspicions of its services or doubts about its honesty. *Mistrust* collocates with words referring to things such as *bank, computer, hospital, profession, system*, etc.[16] As evidence, in a context like *She regards strangers with suspicion*, only the word *distrust* can replace *suspicion*.

5.4.2.5 un- vs. dis-

The negative prefixes *un-* and *dis-* represent different domains. *Un-* evokes the domain of *privation*. It means 'bereft of what is specified by the nominal base'. *Dis-* evokes the domain of *reversal*. It means 'turning around the action signified by the verbal base'. A look at the data in the corpus and the Internet reveals that words annexed to *un-* accompany nouns referring to people or their deeds. By contrast, words annexed to *dis-* accompany nouns referring to people. The prefix *un-* merely reports the absence of the thing denoted in the base, which is due to internal force. The prefix *dis-* opts for the use of external force in carrying out the action of reversal. To see the contrast, let us consider a negative pair:

(24) unqualified vs. disqualified
  a. He was totally *unqualified* for his job as a senior manager.
  b. He was *disqualified* from the competition for using drugs.

The two words are derived from the base *qualified*, meaning 'having finished a training course, or having succeeded in getting into a competition'. Nevertheless, there is a distinction in their use, which rests on the way the base is construed. In (24a), the adjective *unqualified* means 'bereft of the necessary qualifications or requirements'. *An unqualified person* is one who does not have the right knowledge, experience or qualifications to do something. *Unqualified* collocates with words referring to people such *accountant, employee, manager, staff, teacher*, or their deeds such as *challenge, declaration, endorsement, praise, rejection*, etc. In (24b), the verb *disqualify* means 'to prevent somebody from doing something because s/he has broken a rule'. *To disqualify a person* is to pronounce him/her ineligible for an office or activity because of an offence or infringement. *Disqualify* collocates with words referring to people such as *athlete, candidate, competitor, participant, team*, etc. As evidence, only *dis-* implies actions that are beyond one's control, as in *A heart condition disqualified him for military service.*[17]

5.4.2.6 *dis- vs. mis-*

The negative prefixes *dis-* and *mis-* represent different domains. *Dis-* demands the domain of *reversal*. It means 'turning around the action signified by the verbal base'. *Mis-* demands the domain of *treatment*. It means 'falsely doing the action conveyed by the verbal base'. A glance at the data in the corpus and the Internet shows a difference in the behaviour of the two prefixes. Words to which *dis-* is added are used with nouns referring to people, whereas words to which *mis-* is added are used with nouns referring to belongings. To expound the contrast, let us consider a negative pair:

(25) displace vs. misplace
   a. Many people have been *displaced* by the fighting.
   b. I seem to have *misplaced* the keys of the doors.

The two words are derived from the base *place*, meaning 'to put something in a particular place, especially when you do it carefully or deliberately'. Nevertheless, the construal of the base in each verb is different, and so causes a distinction in its use. In (25a), the verb *displace* means 'to move away people from their home to another place'. *To displace people* is to force them out of their homeland or established place. *Displace* collocates with words referring to people such as *children, citizens, families, persons, refugee*, etc. In (25b), the verb *misplace* means 'to put something somewhere and then be unable to find it again, especially for a short time'. *To misplace keys* is to put them in the wrong place. *Misplace* collocates with words referring to belongings such as *envelope, key, cellphone, money, wallet*, etc.[18]

5.5 Summary

In this chapter, I have surveyed two theories which account for utterance meaning. According to the reference theory, meaning equals reference. The meaning of an expression is simply and merely the thing that it identifies or picks out in the world, the so-called *referent*. This seems sensible enough in relation to concrete things, but fails to account for abstract things. One version of this is truth-conditional semantics, where speakers know the meaning of an utterance if they know the conditions under which it would be true. The basic idea is that an utterance like 'Snow is white' is true in English if and only if the stuff referred to by the word *snow* exhibits the property denoted by the word *white*. According to conceptualist theory, the meaning of an utterance is not linked to its truth condition. Rather, it is to be identified with the conceptualisation symbolised by the utterance. It is by virtue of the conceptualisation that the expression can be used to refer to entities in the world. One version of this is construal semantics, where speakers are endowed with the ability to conceive a situation in alternate ways and use different linguistic units to express them. Word pairs may invoke the same conceptual content, yet differ semantically relative to the different construals they represent.

Considering the issues discussed so far, we have reached some significant findings:

1) Negative prefixes form lexical contrasts by attaching to the same bases. Although the same base may admit two or more prefixes, the meanings of the resulting derivatives differ. Each prefix highlights a certain facet of a base, and so shapes the overall meaning of the formation. By determining which prefixes fit which facets, one can precisely define the differences between them. In *unmovable*, the prefix *un-* describes an object that can't physically be moved, whereas in *immovable* the prefix *in-* describes a principle or belief that cannot be moved.

2) Even when word pairs look alike, they are not synonymous. Their alternation spells out semantic differences. As Thomas (1983: 80) rightly explains, 'It is, in fact, hard to find an intensive prefix with no semantic resonance at all. To say that any of these paired words have the same meaning may be to indulge a fondness for linguistic curiosities at the expense of precision'. In such a superficially similar pair like *non-permanent* and *impermanent*, for example, *non-* focuses simply on the class of the thing described, whereas *im-* focuses on its changeability.

3) One useful tool for explaining alternation, which is very common, is construal. In Andrews' (1986: 230) words, 'This shows that linguistic analyses can directly study the structure of meaning in language by turning to the surface forms themselves as information-bearing signs'. The meaning of a complex word is explained in terms of the construal imposed on its base, which is linguistically encoded by the use of a particular prefix. In *a non-aligned country*, *non-* identifies class, whereas in *an unaligned country un-* identifies neutrality.

4) Evidence in support of construal is provided by examples of sentences which demonstrate the uses of the word pairs. The sentences are based mainly on the British National Corpus, shortened for the sake of clarity. To explain differences in meaning, two steps are taken. First, the definitions of the common bases of the pairs are

given. Second, the contributions of the rival prefixes are underlined. Through these examples, it becomes easy to see how related words are used in different ways, and how their uses are conditioned by certain semantic factors.

5) Evidence in support of the meanings established comes from collocation, the tendency of words to occur together in a text. Both common and rare collocates are considered in sense disambiguation. The information provided by collocations helps to allocate specific meanings to the occurrences of the word pairs, thus removing the ambiguity surrounding their use. Collocations of the word *unpractical* indicate that it applies to people, whereas collocations of the word *impractical* indicate that it applies to schemes, proposals or projects.

Notes

1 Examples of other pairs that compete include *non-tonal* vs. *atonal* and *non-sexual* vs. *asexual*. For example, *non-sexual* means 'having no distinction of sex', i.e. sexless, as in *He's angry at her for some completely nonsexual reason.* By contrast, *asexual* means 'without sex or sexual organs', i.e. without interest in sexual relationships, as in *She finds him rather asexual.*

2 Two other rivals here are *antisocial* and *unsocial*. A*ntisocial* means 'opposed to the established society', and refers to one who shuns company and defies or breaks society's rules, as in *If you don't join them in the bar, they will think you are anti-social.* By contrast, *unsocial* means 'falling outside the normal working day and thus socially inconvenient', as in *He works unsocial hours in order to achieve his aims.*

3 Examples of other pairs that contrast include *non-approval* vs. *disapproval, non-organised* vs. *disorganised, non-obedience* vs. *disobedience*, etc. For example, *non-approval* means 'lack of acknowledgement that something is satisfactory', as in *Possible reasons for non-approval run the gamut from the usual to the astounding.* By contrast, *disapproval* means 'refusal to approve what someone has done or is doing', as in *There was a note of disapproval in the teacher's voice.*

4 The same contrast is true of other pairs such as *non-academic* vs. *unacademic, non-believer vs. unbeliever, non-controversial* vs. *uncontroversial, non-scientific* vs. *unscientific, non-productive* vs. *unproductive, non-American* vs. *un-American, non-democratic* vs. *undemocratic, non-Christian* vs. *unchristian, non-aligned* vs. *unaligned, non-person* vs. *unperson, non-man* vs. *unman*, etc. For example, as Algeo (1971: 90) describes, *non-American* is a nationality that is not American. By contrast, *un-American* is behaviour that does not display American customs or traditions, as in *It's practically un-American not to like apple pie.*

5 The same contrast applies to other pairs such as *non-eligible* vs. *ineligible, non-rational* vs. *irrational, non-significant* vs. *insignificant, non-personal* vs.

impersonal, non-finite vs. *infinite, non-action* vs. *inaction*, etc. For example, *non-rational* means 'not based on reason', as in *There is a great deal that is non-rational in modern culture*. By contrast, *irrational* means 'not reasonable or logical', as in *His parents were worried by his increasingly irrational behaviour.*

6 Other pairs that undergo the same contrast are *apolitical* vs. *unpolitical, ahistorical* vs. *unhistorical*, etc. For example, *apolitical* means 'not concerned with or involved in politics', as in *The ombudsman's position is apolitical and independent.* By contrast, *unpolitical* means 'not interested in politics', as in *His rhetoric reached out to unpolitical people.*

7 A third rival is *non-moral*, which means 'not related to moral or ethical considerations', as in *Some of the constraints are non-moral in nature.*

8 Examples of other pairs that differ include *disquiet* vs. *unquiet, disestablished* vs. *unestablished, disassociated/dissociated* vs. *unassociated*, etc. For example, *disassociate* means 'cease or break association with something', as in *He disassociated himself from the remarks*. By contrast, *unassociated* means 'not connected or associated with something', as in *The disease unassociated with current infection.*

9 The same contrast can be identified in other pairs such as *unsubstantial* vs. *insubstantial, unartistic* vs. *inartistic, unexpressive* vs. *inexpressive, unmovable* vs. *immovable, unpractical* vs. *impractical*, etc. For example, *unsubstantial* means 'not having physical substance or material nature', as in *He could no longer live on unsubstantial food*, whereas *insubstantial* means 'lacking firmness, strength or solidity', as in *She presented insubstantial evidence to the court*. With *un-*, the description is confined to things, whereas with *in-* it is confined to qualities. Concerning the difference between *un-* and *in-*, Funk (1971: 376–7) and Aronoff (1976: 127–8) speak in terms of lexicalisation. The *in-* formations are the more lexicalised, e.g. *incomparable*, than the *un-* formations, e.g. *uncomparable*. According to Huddleston & Pullum (2002: 1688–9), a number of bases which accept both *un-* and *in-* either show no difference in meaning, or have preferences for one or the other. Their list includes *advisable, consolable, controllable, distinguishable, elastic, escapable, practical, supportable*, etc.

10 Lehrer (1995: 139) cites another pair that behaves similarly. *Anti-demonstration* is 'an attitude or ideology opposing demonstrations', while *counter-demonstration* is 'an event which takes place in response to some other demonstration'.

11 Another pair that undergoes the same contrast is *misform* vs. *malform. Misformed* means 'mistakenly formed', i.e. has poor form, or poor shape, as in *I don't think he is misformed at all. Malformed* means 'abnormally formed', i.e. has a faulty form or imperfect shape, as in *She gave birth to a malformed child.*

12 In describing the concept [father], for instance, the domain of a kinship network has to be evoked. Although this domain captures an important facet of its meaning, other domains are involved as well. In the domain of physical objects, *father* is a physical being with weight and dimensions. In the domain of living things, *father* is a creature who has a beginning and an end. In the domain of family relations, *father* has a significant role to play. Each of these constitutes a domain against which the word *father* is conceptualised. Each is connected with the other domains, in one way or another. The set of domains which provide the

context for the full understanding of a semantic unit is referred to by Langacker (1991: 147) as *matrix*.

13 A third rival is *inhuman*, which means 'lacking positive human qualities', i.e. warmth and mercy, as in *Prisoners of war were subjected to inhuman treatment*. The same contrast can be identified in other pairs such as *non-science* vs. *anti-science*, *non-matter* vs. *anti-matter*, etc. For example, *non-science* means 'a discipline that is not a science', as in *The distinction between science and non-science is straightforward for the rationalist.* By contrast, *anti-science* means 'a position critical of science and scientific method', as in *The book also moves close to a highly critical analysis of science and yet never becomes anti-science.*

14 The same contrast can be noticed in other pairs such as *unpressed* vs. *depressed*, *unformed* vs. *deformed*, *unfrosted* vs. *defrosted*, etc. For example, *unfrosted* means 'not frosted', as in *The chicken has been left unfrosted*. By contrast, *defrosted* means 'to become no longer frozen', as in *Frozen pasta rarely has to be defrosted before use*.

15 Other pairs that undergo the same contrast are *uninterpreted* vs. *misinterpreted*, *unrecorded* vs. *misrecorded*, *undiagnosed* vs. *misdiagnosed*, etc. For example, *uninterpreted* means 'not interpreted', as in *Some children already have a backlog of uninterpreted experiments*. By contrast, misinterpreted means 'falsely interpreted', as in *His speech has been misinterpreted by the press.*

16 Two other pairs that undergo the same contrast are *disinformation* vs. *misinformation* and *disuse* vs. *misuse*. *Disuse* means 'a situation in which something is not used any more', as in *The old bridge had fallen into disuse*, whereas *misuse* means 'to use something in the wrong way or for the wrong purpose', as in *She had grossly misused her power*. Related to the pair is a third rival, which is *abuse*. *Abuse* means 'to use something for a bad purpose in order to get a personal advantage', as in *The politician has been accused of abusing power*. Likewise, it means 'to treat someone wrongly, especially sexually, physically or emotionally', as in *He was arrested on charges of abusing children*.

17 Two other pairs that undergo the same contrast are *unarmed* vs. *disarmed*, and *uninfected* vs. *disinfected*. For example, *unarmed* means 'not carrying a weapon', as in *He walked into the camp alone and unarmed*, while *disarmed* means 'to take a weapon or weapons away from somebody', as in *Most of the rebels were captured and disarmed*.

18 Another pair that undergoes similar analysis is *dislocate* vs. *mislocate*. *Dislocate* means 'to force something out of its proper position', as *She dislocated her knee falling down some steps*. *Mislocate* means 'to specify a wrong location for something', as in *They mislocated the source of the Nile*.

6 Conclusion

6.0 Overview

This study has been concerned with developing a semantic system that can account for subtle details in the formation of negative expressions in English. The system developed here has yielded good results because it has tackled all aspects of word structure. The main concern of the study has been to illustrate the fact that morphology is first and foremost a matter of meaning. The meaning of a composite word emerges as a result of the interaction between the base and the negative prefix. The specific concern of the study has been to testify to the fact that prefixes have meanings of their own which they contribute to the formations in which they appear. To establish their specificity, prefixes have been treated as rivals and exemplified in word pairs. The structure of this concluding chapter is as follows. Section 1 is an introduction, which focuses on the generalities of the study. The next three sections reveal the particularities of the study. Section 2 describes the category axis of morphology, which shows how the meanings of individual prefixes are represented. Section 3 depicts the domain axis of morphology, which shows the basis on which the prefixes are organised into sets. Section 4 portrays the construal axis of morphology, which shows how occurrences of negative utterances are interpreted. Section 5 is a discussion, which weighs up the advantages and disadvantages of the existing theories in the literature.

6.1 Introduction

This study was about the semantics of negative words in English, derived by means of adding negative prefixes to the initial positions of their bases. To tackle this subject, the study has raised three central questions. The first was: Does a negative prefix exhibit multiple senses, and if so, on what basis are its senses organised? Do the senses derive from a primary sense, and if so, how is the primary sense identified? The second was: Do negative prefixes form semantic sets, and if so, on what basis are they grouped together? Do they represent different facets within the sets, and if so, how do they contrast with one another? The third was: Do pairs of words derived by negative prefixes have different readings, and if so, on what basis are they different? Is the difference supported by evidence, and if so, where does it come from? To answer the questions, the study has proposed a three-pronged system, each of which solves one question. Regarding the first question, the study suggested prototype theory. Concerning the second question, the study proposed domain theory. Relating to the third question, the study recommended construal theory.

To provide a unified account of negative prefixation and avoid the difficulties encountered by rule-centred theories, the study integrated the insights of two linguistic methods: theoretical and empirical. The reason is that the system developed here recognises no division between theoretical and empirical methods. The study has shown how the two methods can coexist to give a detailed description of the way language is used. The theoretical method makes the necessary assumptions, while the empirical method provides the tools to verify them. For the theoretical method, language is subjective knowledge stored in the mind of the speaker, whereas for the empirical method language is usage patterns produced by the speaker. The goal was to present a detailed usage-based analysis of prefixal negation in English within the framework of Cognitive Semantics. In the course of the analysis, the study has invoked some of their key principles. Based on such principles, the study has put forward a number of hypotheses and attempted to check their validity through rigorous argumentation, extensive illustration and empirical evidence.

6.2 The category axis of morphology

The first axis in my semantic system pertained to the syntagmatics of word structure, whereby a bound prefix is combined with a free base. In accordance with this view, a negative prefix forms a complex category of interrelated senses. Given such a view, it becomes clear that a negative prefix is described as having a series of senses with an exemplar, called the *prototype*, as its starting point. The prototype is the most widely used sense. The senses of a prefix are defined in terms of semantic properties characterised by similarities to the prototype. The senses of a prefix, in other words, are structured in terms of their distance from the prototype. The more similar a sense is, the nearer it is to the prototype. The less similar a sense is, the more distant it is from the prototype. The meaning of the prefix is dependent on the base to which it is added. In every group of bases, the prefix displays a different meaning which is part of its polysemous character. The meaning of a complex word is a combination of the semantic properties of both the base and the prefix. The main contribution of the category axis is that it establishes the senses of a negative prefix and provides exact definitions for them. On the basis of these definitions, it becomes feasible to group converging prefixes in sets. This task is taken up in the second axis, which is discussed in the next section.

In Table 1, I summarise the morphological descriptions of the primary negative prefixes. Note that adj. stands for adjective, n. for noun and v. for verb.

Table 1: A morphological sketch of the primary negative prefixes

Prefix	Origin	Function	Base	Gradability
a(n)-	Greek	class-preserving	adj n	gradable/ nongradable
de-	Latin	class-changing/ preserving	n	gradable
dis-	Latin	class-changing/ preserving	adj v n	gradable
in-	Latin	class-preserving	adj n	gradable
non-	Latin	class-preserving	n adj v	nongradable
un-	Old English	class-preserving	adj v	gradable

In Table 2, I summarise the morphological descriptions of the secondary negative prefixes.

Table 2: A morphological sketch of the secondary negative prefixes

Prefix	Origin	Function	Base	Gradability
ab-	Latin	class-preserving	adj v	gradable/ nongradable
anti-	Greek	class-preserving/ changing	n	gradable
contra-	Latin	class-preserving	n adj	gradable
counter-	French	class-preserving	n v	gradable
mal-	Latin	class-preserving	v n adj	gradable
mis-	Old English	class-preserving	v n	gradable

Prefix	Origin	Function	Base	Gradability
pseudo-	Greek	class-preserving	adj n	gradable/ nongradable
quasi-	Latin	class-preserving	adj	gradable
semi-	Latin	class-preserving	adj n	gradable/ nongradable
sub-	Latin	class-preserving	adj n v	gradable/ nongradable
under-	Old English	class-preserving	adj n v	gradable/ nongradable

In Table 3, I summarise the semantic contributions of the primary negative prefixes. Note that ■ represents the prototype, ♦ represents direct extensions and • represents indirect extensions.

Table 3: A semantic sketch of the primary negative prefixes

Prefix	Prototype	Periphery
a(n)-	■'divergent from the quality referred to by the adjectival base', as in *an amoral instigator* ♦'unlike the quality referred to by the adjectival base', as in *an ahistorical phenomenon*	•'without the thing referred to by the adjectival base', as in *acardiac* •'not adhering to the belief referred to by the nominal base', as in *atheist*
de-	■'reversing the action described by the nominal base', as in *decontrol* ♦'removing the thing described by the nominal base', as in *debug* ♦'depriving of the thing described by the nominal base', as in *decontaminate*	•'reducing the thing described by the nominal base', as in *debase* •'analysing the thing described in the nominal base', as in *decode* •'getting off the vehicle described by the nominal base', as in *debus* •'cancelling the thing described in the nominal base', as in *decommission*

Prefix	Prototype	Periphery
dis-	■'the converse of the quality signified by the adjectival base', as in *disloyal* ♦'unwilling to consider the state signified by the verbal base', as in *disbelieve*	•'turning around the action signified by the verbal base', as in *discharge* •'ridding of the thing signified by the nominal base', as in *disarm* •'lacking the thing signified by the nominal base', as in *disbelief*
in-	■'the opposite of the property expressed by the adjectival base', as in *inaccurate* ♦'difficult to perform the process expressed by the adjectival base', as in *incomprehensible*	•'empty of the thing expressed by the nominal base', as in *ineptitude* •'being unable to do the action expressed by the nominal base', as in *inaction*
non-	■'failing to do the action described by the nominal base', as in *non-compliance* ♦'not fulfilling the requirement described by the nominal base', as in *a non-member* ♦'different from the quality described by the adjectival base', as in *a non-aggressive approach*	•'devoid of the characteristics described by nominal base', as in *a non-answer* •'resisting the action described by the verbal base', as in *a non-iron suit*
un-	■'the antithesis of what is specified by the adjectival base', as in *uncooperative* ♦'distinct from what is specified by adjectival base', as in *unofficial* ♦'not subjected to what is specified by adjectival base', as in *undressed*	•'inverting what is specified by the verbal bases', as in *unclose* •'taking away what is specified by the nominal base', as in *unchain* •'bereft of what is specified by the nominal base', as in *unease*

In Table 4, I summarise the semantic contributions of the secondary negative prefixes.

Table 4: A semantic sketch of the secondary negative prefixes

Prefix	Prototype	Periphery
ab-	■ 'move away from the state expressed by the base', as in *abnormal* ♦'locate away from the thing expressed by the base', as in *aboral*	•'wrongly perform the action expressed by the base', as in *abuse* •'release one from the action expressed by the base', as in *absolve*
anti-	■'reacting against the thing named by the nominal base', as in *an anti-abortion campaign* ♦'opposed to the thing named by the base', as in *an anti-communist demonstrator* ♦'displaying the opposite characteristics of the thing named by the nominal base', as in *anti-climactic*	•'preventing the thing named by the nominal base', as in *an anti-bacterial chemical* •'hindering the action named by the nominal base', as in *anti-freeze liquid* •'defending against the weapon named by the nominal base', as in *an anti-aircraft weapon*
contra-	■'in comparison with the thing mentioned in the nominal base', as in *contra-distinction* ♦'posed against the thing mentioned in the nominal base', as in *contra-flow*	•'pitched lower or higher than the thing mentioned in the nominal base', as in *a contrabass*
counter-	■'standing against the thing designated by the nominal base', as in *counter-balance* ♦'facing the action or activity designated by the nominal base', as in *a counter-attack*	•'matching the thing designated by the nominal base', as in *a counter-part* •'duplicating the thing designated by the nominal base', as in *a counter-check*
mal-	■'inappropriately executing the action denoted by the verbal base', as in *maladminister* ♦'inappropriate execution of the thing denoted by the nominal base', as in *maladjustment*	•'not characterised by the thing denoted by the adjectival base', as in *maladroit*
mis-	■'falsely performing the action conveyed by the verbal base', as in *misconceive* ♦'false performance of the thing conveyed by the verbal base', as in *miskick*	•'improper performance of the thing conveyed by the nominal base', as in *misadventure*

Prefix	Prototype	Periphery
pseudo-	■'simulating the thing indicated by the nominal base', as in *a pseudo-bulb* ♦'posing as the thing indicated by the adjectival base', as in *a pseudo-democratic ritual*	•'pretending to be the thing indicated by the nominal base', as in *a pseudo-friend*
quasi-	■'apparently the same as that denoted by the adjectival base', as in *a quasi-autonomous company*	•'having some, but not all of the features of the thing denoted by the base', as in *a quasi-independent commission*
semi-	■ 'half the thing described in the adjectival base', as in *semi-circular* ♦'occurring twice during the period described in the adjectival base', as in *semi-weekly*	•'almost but not entirely the thing described in the adjectival base', as in *semi-conscious* •'just like the thing described in the adjectival base', as in *semi-tropical*
sub-	■'below or beneath the thing named by the nominal base', as in *a submarine* ♦'forming part of the thing named by the nominal base', as in *a sub-class* ♦'almost or nearly the thing named by the adjectival base', as in *a sub-clinical infection*	•'subordinate to the thing named by the nominal base', as in *a sub-dean*
under-	■'below or underneath the thing expressed by the nominal base', as in *the underground* ♦'less in degree or quantity than the thing expressed by the adjectival base', as in *undercooked potatoes*	•'lower in status than the thing expressed by the nominal base', as in *an underachiever*

6.3 The domain axis of morphology

The second axis in my semantic system related to the paradigmatics of word structure, whereby a number of bound prefixes stand for the same concept. With reference to this view, negative prefixes form cognitive domains with respect to which their meanings can be identified. A domain is background knowledge with respect to which lexical items can be characterised. It consists of various facets, each of which deals with a special physical or social experience. Given such a view, it becomes clear that a domain encompasses two types of relationship. One is of similarity where the negative prefixes represent the overall meaning of the domain. Another is of difference where the negative prefixes represent discrete facets of the domain. Primarily, the meaning of a negative prefix can be understood in terms of the particular facet it stands for and the special position it occupies within a domain. Secondarily, the meaning of a negative prefix becomes clear in terms of its relationship to the other prefixes in the domain. The main contribution of the domain axis is that it establishes the precise senses of the negative prefixes and assigns them exact roles in the language. The speaker's choice of a prefix depends on conceptualising a situation. This task is taken up in the third axis, which is discussed in the next section.

In Table 5, I summarise the domains which the primary negative prefixes evoke.

Table 5: The domains evoked by the primary negative prefixes

Domains	Facets	Prefixes	Meanings	Examples
distinction	contradictory	*non-*	describes a choice between two actions	*non-interference*
		a(n)-	describes a choice between two features	*asymmetric*
	contrary	*dis-*	evaluates attitude of people	*disobliging*
		un-	evaluates properties of things	*unclear*
		in-	evaluates properties of situations	*inaccurate*
privation	places/things	*de-*	deprives a place or thing of something	*despoil*
	people	*dis-*	causes someone to lack something	*dishonour*
	objects	*un-*	bereaves an object of something	*unease*
removal	places/things	*de-*	removes something from a place or thing	*defog*
	people	*dis-*	rids someone of something	*dispossess*
	objects	*un-*	takes away an object from something	*unhook*
reversal	places/things	*de-*	reverses a place or thing to its original nature	*desegregate*
	people	*dis-*	turns around the position of someone	*displace*
	objects	*un-*	inverts the direction of an object	*unplug*

In Table 6, I summarise the domains which the secondary negative prefixes evoke.

Table 6: The domains evoked by the secondary negative prefixes

Domains	Facets	Prefixes	Meanings	Examples
degradation	quality	*ab-*	describes something as being lower in standard or nature	*abnormal*
	degree	*sub-*	describes something as being lower in amount or level	*sub-human*
	rank	*under-*	describes someone as being inferior or subordinate	*under-servant*
inadequacy	deception	*pseudo-*	describes someone or something as being false	*pseudo-intellectual*
	similarity	*quasi-*	describes someone or something as being similar	*quasi-judicial*
	incompleteness	*semi-*	describes someone or something as having a thing to some degree	*semi-literate*
opposition	attitude	*anti-*	reacts against a practice	*anti-abortion*
	action	*counter-*	responds to an action	*counter-march*
	comparison	*contra-*	compares one thing to another	*contra-indication*
treatment	accidental	*mis-*	treats an entity unconsciously	*misconceive*
	intentional	*mal-*	treats an entity purposely	*mal-administer*

6.4 The construal axis of morphology

The third axis in my semantic system revolved around the interpretation of negative utterances, whereby pairs of words sharing the same bases and hosting different prefixes are accounted for. With regard to this view, prefixally-negated words are not considered synonymous even if they are morphologically related. The contrast can be shown in terms of construal, a way the speaker conceives and expresses a situation. Given such a view, it becomes clear that even if pairs of negative words have common bases, they display semantic differences relative to the construal the speaker imposes on them. The meaning of a negative word cannot be fully understood without taking into consideration the role of the speaker in shaping it. There are two keys to using a negative pair correctly. One key is to know that the two words constitute different conceptualisations of the same situation. The different conceptualisations reflect different mental experiences of the speaker. The other key is to know that, as a result, the two words are realised morphologically differently. In each morphological realisation, it is the negative prefix that encodes the intended conceptualisation. The main contribution of the construal axis is that it assigns every word a special mission to carry out in language.

6.5 Discussion

The main point of the current study has been descriptive in the sense of synchronically exploring the ways in which complex words are formed by means of negative prefixes. To do this, the study has proposed a new semantic system, in which derivation is seen as the process by means of which a dependent substructure (prefix) is integrated, due to phonological and semantic correspondences, with an autonomous substructure (base) to form a composite structure (new word). Morphology is then the study of the meaning relationships between the components of a complex negative word. A negative word is a symbolic structure, combining a meaning structure with a phonological structure. It consists of two components parts: a base and a prefix. A base is an autonomous phonological structure to which a negative prefix is added. A negative prefix is a phonologically and semantically dependent unit in the sense that it needs the base to complete its meaning. To demonstrate the viability of the new system, it seems inevitable to compare it with the other theory of derivation, i.e. *building-block*, which has been dominant.

On the assumption of the *building-block* theory, a negative word is derived out of its component parts. By applying its principles to complex negative words, one can deduce the following: (1) The subparts of a complex negative word build up its overall form. (2) The segmentation of a complex negative word into discrete parts is primary. (3) The combination of the subparts of a complex negative word is determined by rules. (4) The construction of a complex negative word is carried out independently of the language user, linking elements in accordance with formal parameters. (5) The relationships between the subparts of a complex negative word are described in structural terms or categorial selection. Within the framework of this theory, two approaches of derivation have been singled out in the literature.

According to the *Item-and-Process approach*, which is advocated by Aronoff (1976), Anderson (1992) and Beard (1995), derivation is a set of transformations where a derivational morpheme is added to a base and ultimately modifies its phonological representation. For instance, the negative word *indirect* is the result of the base *direct* and a transformation that adds *in-* to it. In this approach, the lexical base serves as the head of the derivative. As the data have shown, this approach does not provide a convincing account of negative derivation. Because it ignores meaning in the derivation, it is considered ineffective. The transformations do not bear meaning; consequently derivatives sharing the same bases are similar. This inevitably leads to the conclusion that a lexical pair like *uninterpreted* vs. *misinterpreted*, for example, is semantically alike. The surface differences are the result of two different transformations: one transformation adds the prefix *un-*, whereas the other transformation adds the prefix *mis-*.

According to the *Item-and-Arrangement approach*, which is proposed by Williams (1981), Selkirk (1982) and Lieber (1992), derivation is a matter of lexical selection where a derivational morpheme is selected from the lexicon and copied into the appropriate place in a word structure. For instance, to form the negative word *indirect* the affix *in-* is selected from the lexicon and attached to the initial position on its base. In this approach, the prefix serves as the head of the derivative. It assigns grammatical and semantic categories to it. As the data have shown, this approach does not fare any better. It does not provide a clear account which would justify the presence of irregular data. Its inefficiency can be immediately detected when it comes to a lexical pair like *uninterpreted* vs. *misinterpreted*. In such cases, it begs the question of what motivates the occurrence of two prefixes on the same base.

As a reply, a new theory of derivation, i.e. *scaffolding*, has been introduced and applied to the description of prefixal negation. On the assumption of the *scaffolding* theory, a word is derived out of the meanings of its component parts. By applying its principles to complex negative words, one can deduce the following: (1) The subparts of a complex negative word contribute to its overall meaning. (2) The segmentation of a complex negative word into discrete parts is secondary. (3) The combination of the subparts of a complex negative word is sanctioned by schemas. (4) The construction of a complex negative word is carried out by the language user, linking elements in accordance with semantic parameters. (5) The relationships between the subparts of a complex negative word are described in semantic terms or conceptual parameters. Within the framework of this theory, two approaches of derivation have been diagnosed in the literature.

According to the *Morphology-as-Connections approach*, which is posited by Bybee (1985), morphology is represented as connections between the words in the lexicon. This theory is built on inflection, in which it is found that complex words have multiple connections varying in strength. High frequency words undergo less analysis than low frequency words. Low frequency words are stored in terms of their parts. The central assumption of this approach is that a morphological rule reduces to the representation of memorised items in mental storage. However, as the data have shown, the approach lacks in explicitness. Whereas it works for inflectional morphology well, it is not possible to determine its efficacy in accounting for derivational data. It is a still a mystery how this approach can explain the derivation of a pair from a common base, account for the semantic difference between the pair members and provide evidence to support the difference. Above all, does this approach have the right tools to offer plausible solutions for the questions?

According to the *Category-Domain-Construal approach*, which I have postulated in the current study, derivation is a matter of integrating the meanings of the subparts of a complex word. To understand a complex negative word, three pillars seem necessary. (1) The contribution that a given prefix makes to the overall derivative. The multiple senses of a prefix exhibit minimal differences. The prefix is the most important part because it is responsible for the character of the word. (2) The relation of the prefix to the particular facet within a certain concept. Prefixes represent different facets within a domain, and so acquire specific roles in the language. (3) The circumstances under which a given prefix together with the resulting word is used. The meaning of a word is determined by the particular construal imposed on its conceptual content. Therefore, different words code different meanings. Some bases are ambivalent in that they take two or more prefixes. In making the choice, the speaker evaluates the parameters of the speech situation.

To research the nature of the lexicon, my approach hinges upon three crucial arguments, which are in sharp contrast to current lexicographical practices.

The first argument pertains to the phenomenon of *polysemy*, an approach to the lexicon in which one form has many senses. It refers to the capacity of a lexical item to have multiple senses, each of which is applied to a different situation. In *anti-immigration*, *anti-racist* and *anti-hero*, for example, the negative prefix *anti-* is considered a single prefix having three senses. The senses are related and share the concept of opposition. Therefore, they should be treated under one entry in dictionaries. *Polysemy* has been taken as an alternative to the notion of *homonymy*, an approach to the lexicon in which a lexical item shares the same spelling or pronunciation with another item but has a different meaning. In *anti-immigration*, *anti-racist* and *anti-hero*, the negative prefixes are assumed to be homonyms. The senses are unrelated in origin and share no concept. Therefore, they are treated separately in dictionaries. In doing so, dictionaries ignore how senses are related to one another, or how they are motivated. As a result, dictionaries miss the point that the meaning a lexical item has is vital in explaining the peculiarity associated with its behaviour.

The second argument pertains to the phenomenon of *domain*, an approach to the lexicon in which many forms embody one concept. It refers to lexical items which are related in meaning, and so are grouped together into sets. The grouping of the words into sets depends on their core definitions. The negative prefixes *de-* in *despoil*, *dis-* in *disrespect*, and *un-* in *unpeace*, for example, all express the concept of privation, and so should cross-refer to one another in dictionaries. This helps to reveal the specifics of each prefix by first linking it to the appropriate facet and second by contrasting it with its counterparts. *Domain* has been set in contrast to the notion of *synonymy*, an approach to the lexicon in which two or more lexical items tend to express the same meanings. The negative prefixes *de-* in *despoil*, *dis-* in *disrespect*, and *un-* in *unpeace* denote the domain of privation. They are tackled in separate places in dictionaries without any cross reference and without showing their precise meanings. In doing so, dictionaries fail to show that many of these items have something in common as well as an element of difference. As a result, dictionaries lose sight of the exact behaviour of each of the negative prefixes.

The third argument pertains to the phenomenon of *construal*, an approach to the lexicon which takes the role of the speaker in encoding and decoding language. It refers to the way in which a particular situation is perceived in the world and expressed in language. The negative words *uninterpreted* vs. *misinterpreted*, for example, are treated as being unequal in meaning, and so distinctive in use. When the speaker construes something as not being done yet, s/he chooses the word *uninterpreted*. When the speaker construes something as being done but falsely, s/he opts for the word *misinterpreted*. *Construal* has been used as a reply to the notion of *free variation*, an approach to the lexicon in which two or more lexical items may substitute for one another in the same environment without causing a change in meaning. The negative words *uninterpreted* vs. *misinterpreted* are treated as being equal in meaning, and so interchangeable in use. In doing so, dictionaries fail to show that no two words are free variants even if they share the same source. As a result, dictionaries disregard the fact that every lexical item has a certain mission to achieve in discourse.

In brief, in the former theories of morphology words are viewed as the outcome of morphological processes or the product of word-generating rules. In these theories, rules of word formation constitute the exclusive field of investigation. For this reason, little attention has been given to the aspect of meaning in morphological structure. This has, accordingly, rendered such theories unfit for studying the intricacies of prefixal negation. In particular, they lose sight of the differences in meaning between morphologically-related pairs of negative words. Instead of laying emphasis on the procedural aspect of morphological structure, the system proposed here places utmost importance on its semantic aspect. In this study, morphology is looked upon as a system in which various axes of different strengths interact. Only by studying the interaction of these semantic axes can one gain insights into the nature of morphological issues. I hope that other linguists will be able to use the information presented in this study to gain further understanding of the complex interaction between morpho-lexicology and semantics in other languages.

References

Adams, Valerie. 1973. *An Introduction to Modern English Word-formation.* London: Longman.

Adams, Valerie. 2001. *Complex Words in English.* Essex: Longman.

Algeo, John. 1971. 'The voguish uses of *non*'. *American Speech* 46: 87–105.

Anderson, Stephen. 1992. *A-Morphous Morphology*. Cambridge: Cambridge University Press.

Andrews, Edna. 1986. 'A synchronic analysis of *de-* and *un-* in American English'. *American Speech* 61: 221–232.

Aronoff, Mark. 1976. *Word Formation in Generative Grammar*. Cambridge MA: MIT Press.

Aronoff, Mark. 1980. 'The relevance of productivity in a synchronic description of word formation'. In Jacek Fisiak (ed.) *Historical Morphology*: 71–82.

Aronoff, Mark & Kirsten Fudeman. 2005. *What is Morphology?* Oxford: Blackwell.

Atkins, B. T. & B. Levin. 1991. 'Admitting impediments'. In U. Zernick (ed.) *Lexical Acquisition: Exploring on-line resources to build a lexicon.* Hillsdale, NJ: Lawrence Erlbaum.

Baayan, Harald & Rochelle Lieber. 1991. 'Productivity and English derivation: a corpus-based study'. *Linguistics* 29: 801–843.

Bauer, Laurie. 1983. *English Word-formation.* Cambridge: Cambridge University Press.

Bauer, Laurie. 2001. *Morphological Productivity*. Cambridge: Cambridge University Press.

Bauer, Laurie. 2003. *Introducing Linguistic Morphology*. Edinburgh: Edinburgh University Press.

Beard, Robert. 1995. *Lexeme-Morpheme Base Morphology*. New York: State University of New York Press.

Beard, Robert, & Bogdan Szymanek. 1988. *Bibliography of Morphology, 1960–1985*. Amsterdam: Benjamins.

Biber, Douglas et al. 2002. *Longman Student Grammar of Spoken and Written English.* Harlow: Longman.

Biber, Douglas, Susan Conrad & Randi Reppen. 1998. *Corpus Linguistics: Investigating linguistic structure and use.* Cambridge: Cambridge University Press.

Bolinger, Dwight. 1965. 'The atomisation of meaning'. *Language* 41: 555–573.

Bolinger, Dwight. 1977. *Meaning and Form*. London: Longman.

Booij, Geert. 2005. *The Grammar of Words*. Oxford: Oxford University Press.

Brugman, Claudia. 1988. *The Story of* over. *Polysemy, Semantics and the Structure of The Lexicon*. New York: Garland Press.

Brugman, Claudia & George Lakoff. 1988. 'Cognitive topology and lexical networks'. In S. Small, G. Cottrel & M. Tannenhaus (eds) *Lexical Ambiguity Resolution*: 477–507.

Bussman, Hadumod. 1996. *Routledge Dictionary of Language and Linguistics*. London: Routledge.

Bybee, Joan. 1985. *Morphology: An inquiry into the relation between meaning and form*. Amsterdam: Benjamins.

Bybee, Joan & Paul Hopper. 2001. 'Introduction'. In Joan Bybee & Paul Hopper (eds) *Frequency and the Emergence of Linguistic Structure*: 1–24.

Bybee, Joan et al. 1994. *The Evolution of Grammar: Tense, aspect and modality in languages of the world*. Chicago: Chicago University Press.

Carstairs-McCarthy, Andrew. 1992. *Current Morphology*. London: Routledge.

Chomsky, Noam. 1995. *The Minimalist Program*. Cambridge: MIT Press.

Clark, Eve. 1973. 'Non-linguistic strategies and the acquisition of word meanings'. *Cognition* 2: 161–182.

Clark, Eve. 1987. 'The principle of contrast: A constraint on language acquisition'. In B. MacWhinney (ed.) *Mechanisms of Language Acquisition*: 1–33.

Clark, Eve. 1993. *The Lexicon in Acquisition*. Cambridge: Cambridge University Press.

Colen, A. 1980. 'On the distribution of *un-*, *de-* and *dis-* in English verbs expressing reversativity and related concepts'. *Studia Germanica Gandensia* 21: 127–152.

COBUILD. 2001. *Collins Cobuild English Dictionary for Advanced Learners*. Glasgow: HarperCollins.

COBUILD on CD-ROM. 1995. Glasgow: Harper Collins Publishers.

Collins COBUILD *Word Formation*. 1993. London: HarperCollins Publishers.

Croft, William & D. Allen Cruse. 2004. *Cognitive Linguistics*. Cambridge: Cambridge University Press.

Cutler, Anne. 1980. 'Productivity in word formation'. *Chicago Linguistic Society* 16: 45–51.

Cuyckens, Hubert et al. 2003. *Cognitive Approaches to Lexical Semantics*. Berlin: Mouton de Gruyter.

Dressler, Wolfgang. 1985. *Morphology: The dynamics of derivation*. Ann Arbor: Kroma.

Evans, Vyvyan. 2003. *The Structure of Time: Language, meaning and temporal cognition*. Amsterdam: Benjamins.

Evans, Vyvyan & Melanie Green. 2006. *Cognitive Linguistics: An introduction*. Edinburgh: Edinburgh University Press.

Fabb, Nigel. 1988. 'English suffixation is constrained only by selectional restrictions'. *National Language and Linguistic Theory* 6: 527–539.

Fauconnier, Gilles. 1985. *Mental Spaces: Aspects of meaning construction in natural language*. Cambridge: Cambridge University Press.

Fauconnier, Gilles. 1997. *Mappings in Thought and Language*. New York: Cambridge University Press.

Fillmore, Charles. 1977. 'Scenes-and-frames'. In Antonio Zampolli (ed.) *Linguistic Structures Processing*: 55–81.

Fillmore, Charles. 1982. 'Frame Semantics'. In Linguistic Society of Korea (ed.) *Linguistics in the Morning Calm*: 111–138.

Fillmore, Charles & Beryl Atkins. 1992. 'Towards a frame-based lexicon: the semantics of *risk* and its neighbours'. In A. Lehrer & E. Feder Kittay (eds) *Frames, Fields, and Contrasts: New essays in semantics and lexical organisation.* Hillsdale, NJ: Lawrence Erlbaum.

Fillmore, Charles. & Beryl Atkins. 2002. 'Describing polysemy: the case of *crawl*'. In Yael Ravin & Claudia Leacock (eds) Polysemy: *Theoretical and computational approaches*: 91–110.

Funk, Wolf-Peter. 1971. 'Adjectives with negative affixes in modern English and the problem of synonymy'. *Zeitschrift für Anglistik und Amerikanistik* 19: 364–386.

Geeraerts, Dirk. 1988. 'Cognitive grammar and the history of lexical semantics'. In Brygida Rudzka-Ostyn (ed.) *Topics in Cognitive Linguistics*: 647–677.

Givón, Talmy. 1990. *Syntax: a functional-typological introduction.* Amsterdam: Benjamins.

Goddard, Cliff. 1998. *Semantic Analysis: A practical introduction.* Oxford: Oxford University Press.

Goldberg, Adele. 1996. *Conceptual Structure, Discourse and Language.* Stanford: CSLI Publications.

Greenbaum, Sidney & Janet Whitcut. 1988. *Longman Guide to English Usage.* Essex: Longman.

Halle, M. 1973. 'Prolegomena to word formation'. *Linguistic Inquiry* 4: 3–16.

Hamawand, Zeki. 2007. *Suffixal Rivalry in Adjective Formation. A cognitive-corpus analysis.* London: Equinox.

Hamawand, Zeki. 2008. *Morpho-Lexical Alternation in Noun Formation.* London: Palgrave Macmillan.

Haspelmath, Martin. 2002. *Understanding Morphology*. London: Arnold.

Heine, Bernd. 1997. *Cognitive Foundations of Grammar*. Oxford. Oxford University Press.

Hockett, Charles. 1954. 'Two models of grammatical description'. *Word* 10: 210–231.

Hopper, Paul & Elizabeth Traugott. 1993. *Grammaticalisation.* Cambridge: Cambridge University Press.

Horn, Laurence. 1989. *A Natural History of Negation.* Chicago: University of Chicago Press.

Horn, Laurence. 2002. 'Uncovering the un-word: a study in lexical pragmatics'. *Sophia Linguistica* 49: 1–64.

Huddleston, Rodney & Geoffrey Pullum. 2002. *The Cambridge Grammar of the English Language*. Cambridge: Cambridge University Press.

Jackendoff, Ray. 1983. *Semantics and Cognition.* Cambridge: MIT Press.

Jackendoff, Ray. 1990. *Semantic Structures*. Cambridge: MIT Press

Jackendoff, Ray. 1997. *The Architecture of the Language Faculty*. Cambridge: MIT Press.

Jensen, John. 1990. *Morphology. Word Structure in Generative Grammar*. Current issues in linguistic theory 70. Amsterdam: Benjamins.

Jewell, Daniel. 2001. *The negative adjectival prefix in English*. Unpublished M.A. Thesis. Brigham Young University.

Johnson, Mark. 1987. *The Body in the Mind: The bodily basis of meaning, imagination, and reason*. Chicago: University of Chicago Press.

Katamba, Francis. 1993. *Morphology*. London: MacMillan Press Ltd.

Katz, Jerrold. 1972. *Semantic Theory*. New York: Harper & Row.

Katz, Jerrold. & Jerry Fodor. 1963. 'The structure of a semantic theory'. *Language* 39: 170–210.

Kemmer, Suzanne & Michael Barlow. 2000. 'Introduction'. In Michael Barlow & Suzanne Kemmer (eds) *Usage-Based Models of Language*: vii-xxviii.

Kennedy, Graeme. 1991. 'Between and through: The company they keep and the functions they serve'. In Karin Aijmer & Bengt Altenberg (eds) *English Corpus Linguistics*: *Studies in honour of Jan Svartvik*: 95–110.

Kilgarriff, Adam & David Tugwell. 2001. 'Word sketch: extraction and display of significant collocations for lexicography'. In Proceedings of ACL Workshop on Collocation: Computational Extraction, Analysis and Exploitation. Toulouse, France: 32–38.

Lakoff, George. 1987. *Women, Fire and Dangerous Things: What categories reveal about the mind*. Chicago: University of Chicago Press.

Lakoff, George. 1990. 'The invariance hypothesis: is abstract reason based on image schemas?' *Cognitive Linguistics* 1: 39–74.

Lakoff, George & Mark Johnson. 1980. *Metaphors We Live By*. Chicago: University of Chicago Press.

Langacker, Ronald. 1987. *Foundations of Cognitive Grammar.* Vol.1: *Theoretical Prerequisites*. Stanford: Stanford University Press.

Langacker, Ronald. 1988a. 'A view of linguistic semantics'. In Brygida Rudzka-Ostyn (ed.) *Topics in Cognitive Linguistics*: 49–90.

Langacker, Ronald. 1988b. 'A usage-based model'. In Brygida Rudzka-Ostyn (ed.) *Topics in Cognitive Linguistics: Current issues in linguistic theory*: 127–161.

Langacker, Ronald. 1991. *Foundations of Cognitive Grammar.* Vol. 2: *Descriptive Application*. Stanford: Stanford University Press.

Langacker, Ronald. 1994. 'Cognitive Grammar'. In R. E. Asher (ed.) *The Encyclopedia of Language and Linguistics* 2: 590–594.

Langacker, Ronald. 1997. 'The contextual basis of cognitive semantics'. In Jan Nuyts & Eric Pederson. *Language and Conceptualisation*: 229–232.

Langacker, Ronald. 2000. 'A dynamic usage-based model'. In Suzanne Kemmer & Michael Barlow (eds) *Usage Based Models of Language*: 1–63.

LDOCE. 2003. *Longman Dictionary of Contemporary English*. Harlow: Longman.

Leech, Geoffrey 1981. *Semantics: The study of meaning*. Harmondsworth: Penguin.

Lees, Robert. 1960. *The Grammar of English Nominalisations*. The Hague: Mouton.

Lehrer, Adrienne. 1974. *Semantic Fields and Lexical Structure*. Amsterdam, North- Holland.

Lehrer, Adrienne. 1978. 'Structures of the lexicon, and transfer of meaning'. *Lingua* 45: 95–123.

Lehrer, Adrienne. 1995. 'Prefixes in English word formation'. *Folia Linguistica* 29.1–2: 133–148.

Lehrer, Adrienne & Eva Feder Kittay (eds). 1992. *Frames, Fields and Contrasts: New essays in semantic and lexical organisation*. Hillsdale, New Jersey: Lawrence Erlbaum.

Levin, Beth. 1993. *English Verb Classes and Alternations: A preliminary investigation*. Chicago: University of Chicago Press.

Lieber, Rochelle. 1992. *Deconstructing Morphology: Word formation in syntactic theory*. Chicago: University of Chicago Press.

Lieber, Rochelle. 2004. *Morphology and Lexical Semantics*. Cambridge: Cambridge University Press.

Lieber, Rochelle. 2005. 'English word-formation processes'. In Pavol Stekauer & Rochelle Lieber. *Handbook of Word-Formation*: 375–427.

Longman Language Activator. 2002. Pearson Education Limited.

Lyons, John. 1977. *Semantics*. Vols. 1 & 2. Cambridge: Cambridge University Press.

Lyons, John. 1995. *Linguistic Semantics: An introduction*. Cambridge: Cambridge University Press.

Marchand, Hans. 1969. *The Categories and Types of Present-day English Word Formation. A Synchronic-Diachronic Approach*. Munich: Beck.

Matthews, Peter. 1972. *Inflectional Morphology*. Cambridge: Cambridge University Press.

Matthews, Peter. 1974. *Morphology*. Cambridge: Cambridge University Press.

Maynor, Natalie. 1979. 'The morpheme *un-*'. *American Speech* 54: 310–311.

Merriam-Webster Dictionary Online. Available at: http://www.m-w.com.

Nerlich, Brigitte et al. 2003. *Polysemy: Flexible patterns of meaning in mind and language*. Berlin: Mouton de Gruyter.

OALD. 2005. *Oxford Advanced Learner's Dictionary*. Oxford: Oxford University Press.

Oxford English Dictionary on CD-ROM. 1994. Oxford: Oxford University Press.

Peters, Pam. 1995. *The Cambridge Australian English Style Guide*. Cambridge: Cambridge University Press.

Peters, Pam. 2004. *The Cambridge Guide to English Usage*. Cambridge: Cambridge University Press.

Partridge, Eric. 1961. *Usage and Abusage. A guide to good English.* London: Hamish Hamilton.
Plag, Ingo. 1999. *Morphological Productivity. Structural Constraints in English Derivation.* Berlin: Mouton de Gruyter.
Plag, Ingo. 2003. *Word-Formation in English.* Cambridge: Cambridge University Press.
Pustejovsky, James. 1995. *The Generative Lexicon.* Cambridge: MIT Press.
Quirk, Randolph et al. 1985. *A Comprehensive Grammar of the English Language.* London. Longman.
Ravin, Yael & Claudia Leacock. 2002. *Polysemy: An overview.* Oxford: Oxford University Press.
Riemer, Nick. 2005. *The Semantics of Polysemy: Readings meaning in English and Warlpiri.* Berlin: Mouton de Gruyter.
Rosch, Eleanor. 1977. 'Human categorisation'. In Neil Warren (ed.) *Studies in Cross-Cultural Psychology* 1: 3–49.
Rosch, Eleanor. 1978. 'Principles of categorisation'. In E. Rosch & B. B. LIoyd (eds) *Cognition and Categorisation*: 27–48.
Ruhl, Charles. 1989. *On Monosemy: A study in linguistic semantics.* SUNY Series in Linguistics: Albany: State University of New York Press.
Ryder, Mary Ellen. 1994. *Ordered Chaos: The interpretation of English noun-noun compounds.* Berkeley: University of California Press.
Selkirk, Elisabeth. 1982. *The Syntax of Words.* Cambridge MA: MIT Press.
Spencer, Andrew. 1991. *Morphological Theory.* Oxford: Blackwell.
Stekauer, Pavol. 2000. *English Word-formation. A History of Research (1960–1995).* Tübingen: Gunter Narr Verlag.
Stump, Gregory. 2001. *Inflectional Morphology. A Theory of Paradigm Structure.* Cambridge: Cambridge University Press.
Sweetser, Eve. 1990. *From Etymology to Pragmatics: Metaphorical and cultural aspects of semantic structure.* Cambridge: Cambridge University Press.
Szymanek, Bogdan. 1998. *Introduction to Morphological Analysis.* Warszawa: Wydawinctwo Naukowe PWN.
Talmy, Leonard. 1983. 'How language structures space'. In Herbert Pick & Linda Arcedolo (eds) *Spatial Orientation: Theory, research, and application*: 225-282.
Talmy, Leonard. 1985. 'Force dynamics in language and thought'. *Chicago Linguistic Society* 21:293–337.
Taylor, John. 1989. *Linguistic Categorisation. Prototypes in Linguistic Theory.* Oxford: Clarendon Press.
Taylor, John. 2002. *Cognitive Grammar.* Oxford: Oxford University Press.
Taylor, John & Robert E. MacLaury. 1995. *Language and the Cognitive Construal of the World.* Berlin: Mouton de Gruyter.
Thomas, Macklin. 1983. 'Interchangeable pairs in un-, in-, en, etc'. *American Speech* 58: 78–80.

Tomasello, Michael. 2000. 'First steps toward a usage-based theory of language acquisition'. *Cognitive Linguistics* 11: 61–82.

Tottie, Gunnel. 1980. 'Affixal and non-affixal negation in English: two systems in (almost) complementary distribution'. *Studia Linguistica* 34: 101–123.

Trier, Jost. 1931. *Der deutsche Wortschatz im Sinnbezirk des Verstandes : die Geschichte eines sprachlichen Feldes*. Heidelberg: Winter.

Tuggy, David. 2005. 'Cognitive approach to word formation'. In P. Stekauer & R. Lieber (eds) *Handbook of Word-Formation*: 233–265.

Tummers, Jose et al. 2005. 'Usage-based approaches in cognitive linguistics: A technical state of the art'. *Corpus Linguistics and Linguistic Theory* 1–2: 225–261.

Tyler, Andrea. & Vyvyan Evans. 2001. 'Reconsidering prepositional polysemy networks. The case of *over*'. *Language* 77: 724–765.

Urdang, Laurence. 1982. *Suffixes and Other Word-final Elements of English*. Detroit: Gale Research Company.

Welte, Werner. 1978. *Negationslinguistik*. München: Wilhem Fink Verlag.

Wierzbicka, Anna. 1988. *The Semantics of Grammar*. Amsterdam: Benjamins.

Williams, Edwin. 1981. 'Argument structure and morphology'. *Linguistic Review* 1: 81–114.

Williams, Geoffrey. 2002. 'In search of representativity in specialised corpora: Categorisation through collocation'. *International Journal of Corpus Linguistics* 7: 43–64.

Wittgenstein, Ludwig. 1953. *Philosophical Investigations*. Oxford: Basil Blackwell & Mott.

Wittgenstein, Ludwig. 1958. *The Blue and Brown Books*. Oxford: Basil Blackwell & Mott.

Zimmer, Karl. 1964. *Affixal Negation in English and Other Languages*. *Word* 20. Supplement Monograph 5.

Indexes

Linguistic terms

Negative prefixes

Word pairs

www.ingramcontent.com/pod-product-compliance
Lightning Source LLC
LaVergne TN
LVHW010448080826
844660LV00027B/1234

* 9 7 8 1 8 4 5 5 3 9 7 1 9 *